INDUSTRIAL DEMOCRACY

SOME OTHER VOLUMES IN THE
SAGE FOCUS EDITIONS

INDUSTRIAL DEMOCRACY

Strategies for Community Revitalization

Edited by
Warner Woodworth
Christopher Meek
William Foote Whyte

SAGE PUBLICATIONS
Beverly Hills   London   New Delhi

For information address:

SAGE Publications, Inc.
275 South Beverly Drive
Beverly Hills, California 90212

SAGE Publications India Pvt. Ltd.
M-32 Market
Greater Kailash I
New Delhi 110 048 India

SAGE Publications Ltd
28 Banner Street
London EC1Y 8QE
England

Printed in the United States of America

Library of Congress Cataloging in Publication Data

Main entry under title:

Industrial democracy.

(Sage focus editions ; v. 73)
Includes index.
1. Urban economics.   2. Community development,
Urban—United States.   3. Industry—Social aspects—
United States.   4. Employee ownership—United States.
5. Urban renewal—United States.   I. Woodworth, Warner.
II. Meek, Christopher.   III. Whyte, William Foote,
1914-   .
HT321.I53   1985   338.973'009173'2   85-11741
ISBN 0-8039-2476-3
ISBN 0-8039-2477-1 (pbk.)

FIRST PRINTING

CONTENTS

FOREWORD

Over the past dozen years, I have sponsored a number of bills designed to encourage the use of Employee Stock Ownership Plans (ESOPs) as a technique of corporate finance. I am happy to report that Congress has found the concept sufficiently attractive to approve several tax incentives designed to promote employee stock ownership in the American workplace.

It is my hope that this idea has now taken root and that in future Congresses we will see a continued expansion of policies that encourage Americans to own a stake in the enterprise in which they spend a good part of their working lives. I am convinced this is an essential component of the very fabric of American society and I am encouraged by the widespread use of ESOPs in a variety of circumstances.

Those contributing to this book bring to the issue of community economic revitalization a wealth of experience and concern. This is a difficult and perplexing problem, and one for which no one person has all the right answers. Yet each of these authors recognizes the potential of labor-management committees and worker ownership as common themes around which a workplace and, indeed, a community can rally in formulating a strategy for economic revitalization.

The evidence indicates that employee stock ownership has the potential for creating a work environment and a reward system that can bring out the best that both the American worker and the American workplace have to offer. It can also lay the groundwork for a style of management based more on commitment than control. Commitment is dependent upon a sense of participation. However, for the most part, the opportunity for economic participation has thus far generally been limited to jobs alone. Worker ownership through ESOPs provides an opportunity to participate in capital ownership.

THE BEST USE OF
ECONOMIC DEVELOPMENT FUNDS

In terms of motivation, the merits of a worker-ownership community revitalization policy can be summarized easily: Ownership counts. It summons up a common determination to succeed and ensures that the company's success is shared with those on whom that success will largely depend.

The increased use of ESOP financing for development financing reflects a return to incentive economics. The coupling of development programs with worker ownership can result in a more production-oriented work force. That, in turn, helps to ensure that scarce economic revitalization dollars are invested so as to get the most "bang for the buck."

In addition to ensuring that the benefits of economic revitalization financing are more widely dispersed than through traditional means, ESOP financing also helps to create the circumstances in which assisted companies are better able to survive and in which taxpayer-provided development loans are more likely to be repaid.

An encouraging and growing trend is reflected in the number of U.S. companies experimenting with labor-management committees, participative management, quality circles, and other programs designed to enhance the quality of work life. Employee stock ownership can help create an environment in which companies are more likely to establish such programs. These programs can have a beneficial effect not only on productivity but also on job satisfaction, individual dignity, community cohesiveness, and general mental health.

COMMUNITY-RESPONSIVE ECONOMIC
DEVELOPMENT STRATEGIES

One of the greatest failings of our free-enterprise system is its hostility to the marginally profitable company. Employee ownership of such firms brings with it a way to correct that glaring deficiency, a deficiency that has devastated entire regions as marginally successful companies all across the nation have closed their doors.

As any free-market economist will tell you, it is the essence of capitalism to allow—indeed, to encourage—financial capital to seek its highest return. It is this "invisible hand" that serves as the driving force of a market economy. By that measure, plant closings make perfectly good economic sense, particularly to those who own capital and to those financial managers hired to oversee that capital on their behalf.

But to those who neither own nor manage—to the workers and the communities who must live with the effects of such closings—the fact that market signals are being followed is faint comfort, particularly when they must cope with swollen welfare rolls and deficits, overburdened state and local relief efforts, increased health and emotional problems, and other repercussions, including the ripple effects of additional unemployment resulting from generally weaker consumer demand.

This is not to suggest that ESOP financing should be utilized in all cases, or in companies that cannot be made profitable. Quite the contrary; a company that is not market responsive, a company that cannot meet its competition and turn a profit, should not put its employees in the position of owning that company. ESOP-type financing is not intended for losers.

It *is* intended, however, for those losers and for those marginally profitable firms that, with employee participation in ownership, can become winners. It is intended for those willing to sacrifice now in order to work for a brighter future.

NEW OPTIONS, OPPORTUNITIES, AND OUTCOMES

Employee stock ownership brings a healthy new dimension to the community economic revitalization process. It enables those affected to examine the issue in a different frame of reference, one in which the effect on them becomes much more real, and one in which they can realize how important their contribution is to the company's success.

Employee stock ownership is for those who believe in the worth of the individual worker and who see the rank and file as the root source of quality, creativity, and productivity. In short, employee stock ownership is intended for those employers and employees who are ready for a new and renewed American workplace, and a new way of thinking, and for those who are prepared for a new array of options, opportunities, and outcomes.

Yet there is much more to the concept of ESOP financing than community economic revitalization. If we in this nation continue to rely solely on traditional techniques of corporate finance, those techniques will continue to concentrate capital ownership further into an already overly concentrated ownership pattern. That would show a great failure of foresight on our part, because not only will we continue to have an inequitable form of free enterprise in the United States, we will also have a form unsuitable for imitation abroad.

This nation needs a more hopeful model—a working model of what we would advocate for other nations. We need to show people all over the world how the increasing prosperity of our private property economy spreads out and reaches Americans in all walks of life.

I urge you to get involved in this crucially important debate. A good way to start is to begin to transform the American workplace. I challenge you to begin the process of making it more humane, more people oriented, and more "us" oriented.

Just as a stream can rise no higher than its source, the American workplace can only be as good as the ideas and ideals that are brought to it. In searching for strategies for community economic revitalization, seek to make the workplace as good as you would have this country be. Dare to be a showcase and others will follow your example.

What employee stock ownership legislation suggests is not a destination, but a new direction, a direction I am convinced the U.S. economy must take if we are to be true to our ideals and realize our full potential as a nation.

—Senator Russell B. Long

PREFACE

The devastating impact of industrial decline in communities through-out the United States, a crisis of major proportions that began in the early 1980s is the subject of this book. Our objective is to explicate the newly developed field of theory and practice for countering this industrial decline and subsequent deterioration of the quality of community life. We begin with an overview of the problem of economic dislocation, followed by an exploration of the social responsibility of corporations to surrounding communities. Analytical contributions by selected authors then survey several models for coping with economic decay. The final section of the book compares these alternative strategies according to critical variables, assesses their impact, and offers an intriguing paradigm for future research and social change.

This book will be of major interest to several groups. (1) Those who are teaching and doing research in such disciplines as sociology, economics, community psychology, labor relations, and management should find this book useful both as a reference and a textbook that offers an introduction to the potential of industrial democracy and social change for upper-division or graduate-level courses. (2) The book will also serve the market of practitioners working on the problems of socioeconomic depression. It should prove to be a valuable resource for labor leaders, managers, and interested parties of the public sector, including city planners and state and federal economic development officials.

—Warner Woodworth
Christopher Meek
William Foote Whyte

PART I

THE PROBLEMS OF ECONOMIC DISLOCATION
An Overview

The five chapters in Part I outline the basic problems to be addressed in subsequent sections of this book. William Foote Whyte begins by introducing a new pattern for local economic development in the United States, an approach that stands in marked contrast to traditional strategies. The new strategy emphasizes grass-roots initiatives and regional resources rather than intervention from above, at the more macro level.

David Moberg graphically captures the plight of the victims of plant closings: workers, their families, and their communities. The financial, social, and emotional costs of conglomerate interests that supersede community needs are reported as the author reviews the flight of main-line industry from the Northeast and Midwest.

Staughton Lynd, a historian and attorney, explores in Chapter 3 the advantages of a "brownfield" model of reindustrialization over the "greenfield" approach of most community economic developers. Suggesting that "big steel's" complaints about obsolescence and non-competitiveness are largely rhetorical, Lynd advocates his view that the costs to industry and communities are overwhelmingly in favor of brownfield modernization, and that, when coupled with ethical and political values about the good society, the choice is clear.

In Chapter 4, Bennett Harrison and Barry Bluestone advance their hotly debated position about why plants close and the corporate logic behind decisions to shut down operations and abandon communities. The "capital shift" theory advanced by the authors is based on important data about disinvestment and deindustrialization, data the authors have amassed not only to pinpoint the causes of plant shutdowns but also to debunk the standard myths that defend executive decision making.

In Chapter 5, Meek and Woodworth trace the problems of community economic disintegration to the issue of absentee ownership. They point out that in contrast to historical patterns of solidarity among local entrepreneurs, workers, and communities, today's business climate has broken up these formerly unitary relationships. Control has slipped from community hands and shifted to the executive boardroom, often thousands of miles away from the community in question. The authors conclude by laying the groundwork for the next two sections of the book, introducing the strategies of labor-management cooperation and worker ownership.

1

NEW APPROACHES TO INDUSTRIAL DEVELOPMENT AND COMMUNITY DEVELOPMENT

William Foote Whyte

A new pattern of organizations and strategies for local economic development is emerging in the United States. This pattern places the emphasis upon local and area initiative and resources, stimulated and supported by the local, state, and federal governments. Because the principal actors in these development strategies are still learning their new roles, there has naturally been much confusion and waste in the process; yet we are seeing here and there solid and even sometimes spectacular achievements. It is our purpose to show the potential of this new pattern through examples of successful development projects achieved by local people, with the essential assistance of government officials, from the village level to Washington.

THE OLD AND
THE NEW PATTERNS

In traditional terms, decentralization means passing down authority and responsibility from the national government to lower levels of government—and the lower the better. As we see it, successful local development projects will continue to depend upon actions taken at national and state levels, yet these actions involve a major shift in the roles of these higher levels of government. Instead of doing things for and to local people, federal and state officials are beginning to devise programs designed to help local people to do things for themselves. In this new framework, successful projects are those that achieve a skillful combination of federal and state support, and guidance with local initiative and resourcefulness.

The new pattern also involves a major shift in thinking about the process of creating jobs. In traditional thinking, the responsibility for job creation rests fundamentally with the private sector. Government can help only by providing infrastructure and public services, and by easing the tax burden on existing firms or on firms that are lured to establish themselves in a particular governmental jurisdiction. The profit motive is considered the principal if not the only spur to economic progress, and therefore government should do only those things that help the private firm maximize profits. High profits lead to high investments, and high investments create jobs.

The emerging new pattern is based upon the implicit assumption that the old pattern is not working. In the first place, the traditional organs of government, from central to local, do not seem to be doing a very effective job in creating the conditions for the growth of employment. At the local level, the traditional development strategy has involved cutthroat competition between one state and another or one municipality and another as each tries to offer incentives that will attract new plants. Although each state may be forced into competition in order to protect itself against raids by other states, clearly such strategies yield no overall benefit to the national economy.

Increasingly, critics have come to question profit maximization as a guiding principle for economic development. This is not to question the importance of profits per se, but rather to distinguish between profit maximization and profits as a limiting condition. If a firm is to survive in the long run without government subsidy, then obviously it must earn enough income not only to cover its operating expenses but also to build reserves for investment so as to remain competitive; but this is not the same as saying that investment should necessarily go to organizations that will yield the highest profit.

Consider the example of Bates Fabrics Company in Lewiston, Maine, an employer of 1,100 workers. Over the years, the parent company had grown into a conglomerate with steadily expanding investments outside of textiles. Conglomerate decision makers found they could make 15 percent to 20 percent on invested capital in energy and natural resources, whereas their last remaining textile operation could expect to make only 5 percent to 7 percent. The profit maximization principle therefore seemed to justify top management's decision to close the Lewiston plant.

If cost and benefits are examined from the community standpoint, the profit question takes quite a different form. In the crisis, the local management, union leaders, and community people came together to obtain a loan guarantee from the Farmers Home Administration to convert the plant (gradually) to employee ownership and maintain the 1,100 jobs. The loan also provided the capital necessary to modernize the plant and keep it competitive.

As this case indicates, a plant can be economically viable while falling far short of meeting the goal of profit maximization for private investors. In today's economy, investors who are concerned simply with profit maximization would have no interest in a firm that promised to yield only 5 percent to 7 percent—and with a considerable risk. The workers in the plant and the people in the community of Lewiston had a substantial interest in keeping that plant operating, so they could not afford to leave their fate in the hands of private investors.

If you think in terms of profit maximization, you are unlikely to find much worth investing in in a depressed community. But if you think of profit as a limiting factor, many possibilities appear.

FEDERAL AND STATE ASSISTANCE

Although this chapter is focused primarily upon local-level projects, it is important to look first at the *enabling* functions of higher levels of government that make these local initiatives possible. A full description of organizations and programs at state and federal levels designed to support local development is beyond the scope of this chapter. Let me simply note those federal and state programs that carried out enabling functions for the local projects I shall describe.

The Economic Development Administration (EDA) program was established by the Public Works and Economic Development Act of 1965. The act aimed to provide special assistance to areas of the country suffering from high unemployment and slow economic growth, stagnation, or decline. For this purpose, EDA grouped together three to five counties in similarly depressed condition and, in collaboration with the county governments, established Economic Development Districts. Although EDA provides the major financial support at the outset, the counties contribute to the financing even at this stage, and, as time goes on, it is expected that counties, cities, and towns will provide steadily increasing proportions of the Economic Development District (EDD) budget.

In a complementary program, EDA has established more than twenty University Technical Assistance Centers, providing 75 percent of the first-year budgets and gradually scaling down the federal contribution to the point at which the centers are financed fully by university, state, or other non-EDA funds. The centers are designed to create new employment opportunities through special studies and consultation with business and industry, local public officials, and economic development districts. The centers also provide vehicles through which students and professors can gain field experience working on the practical development problems of their area, thus strengthening the university as a teaching-learning institution while providing public service that would otherwise be available only through private consultants. Given that in an economically depressed area few small businessmen or local governments could afford to hire private consultants, this EDA program is designed to fill an important gap in information and expertise necessary for local economic development.

The Community Development Corporation (CDC) came into prominence through legislation supporting the Office of Economic Opportunity (OEO). In response to political controversies stirred by some of its programs, President Nixon seemed determined to phase OEO out of existence. Nevertheless, CDCs continue to operate here and there throughout the country, and some of them have performed impressively enough to deserve further public attention and support.

According to the Community Self-Determination Act of 1968, the purposes of the CDC, for urban or rural communities, are as follows:

> expanding their economic and educational opportunities, increasing their ownership of productive capital and property, improving their health, safety, and independence, expanding their opportunities for meaningful decision making, and generally securing the economic development and social well-being and stability of their community area.

At present, the Community Services Administration funds about fifty CDCs in its Title VII program. Other federal support comes from the Community Development Block Grant program administered by cities and supported by allocations from the Department of Housing and Urban Development. However, the great majority of CDCs receive no federal assistance and must support themselves from other sources.

While EDDs are established through agreements between EDA and county officials, CDCs are formed through the initiative of citizens. Building on a base of popular participation, the CDCs' success must also depend upon their ability to secure the support of public officials and to finance themselves through some combination of government and foundation grants and grass-roots fundraising.

Although it has recently been dismembered, with some of its programs distributed to older federal agencies, during its brief existence the National Center for Productivity and Quality of Working Life played a major role in promoting the establishment of city or area labor-management committees. In collaboration with the Federal Mediation and Conciliation Service, the National Center served mainly to provide stimuli and guidance, because it lacked the budget to provide major financial support to such committees.

Although each state has its own organization and development strategy, the general line of activity can be illustrated by the Job Development Authority (JDA) of New York State. JDA issues Industrial Development Revenue Bonds that provide low-interest, long-term financing for industrial development. JDA then loans these funds for construction of new industrial plants, pollution control, buildings for research and development, or rehabilitation of old industrial plants. This program is supported by a familiar pattern of tax incentives.

In New York State a number of counties have established semiautonomous Industrial Development Agencies (IDAs), which can receive and disburse funds from county, state, federal, or private sources. Furthermore, IDAs are empowered to raise their own money through issuing tax-exempt bonds and they may acquire, hold, and dispose of property for economic development purposes. IDAs also have the mandate to conduct studies for planning purposes and to provide technical assistance to development projects.

AN EXAMPLE: MOHAWK VALLEY
ECONOMIC DEVELOPMENT DISTRICT

To illustrate the potential of the EDD, let us look at the Mohawk Valley case. Hardly a typical case, by any standard the performance of MVEDD would rank it as among the most effective in the country.

Nevertheless, it is a good example to use to indicate the potential for such an organization.

MVEDD was created early in the EDA program by a large group of prominent citizens led by John Ladd, an entrepreneur with a background of success in the private sector. Ladd had learned about the possibilities of the new legislation and set about putting together an organization that would win the support of the various counties. In the succeeding years, MVEDD has performed a number of the traditional services, such as helping towns and cities to get federal funds for sewage and for construction of industrial parks. Of more interest here are its innovative activities. I will report briefly on two examples.

An old textile mill in the Mohawk Valley was about to be shut down, putting 260 workers out of their jobs. John Ladd had been working to bring together a group of private investors to buy the plant and keep it operating, but he was unable to put the deal together before the plant was to be sold at auction by Sperry-Rand. Ladd assumed that his principal competitors would be liquidators or, as he called them, "junk dealers," who would bid for the plant in order to strip it of its machines, leaving it of no value except as a warehouse.

The $1.3 million bid of Ladd's group proved to be just $50,000 higher than the next highest bidder, who was indeed a liquidator. Having staved off the stripping of the plant, Ladd now managed to complete negotiations to sell it to a new group of private investors, who are keeping the plant operating and maintaining those 260 jobs. In this case, the resourcefulness of the directors of the EDD and $50,000 made the difference between 260 jobs and an empty building.

The saving of the Library Bureau is a more complex story, but it deserves to be recounted in some detail as it illustrates the variety and complexity of measures that must be taken by the public-sector entrepreneur if jobs are to be saved in a crisis situation. This Herkimer plant had been producing high-quality library furniture since before the turn of the century. In 1955, it was taken over by Sperry-Rand. In the succeeding twenty years, the Herkimer plant yielded a profit every year but one. Nevertheless, Sperry-Rand decided in 1976 to close it down, declare a tax loss, and sell the valuable machinery. Although the plant had been making money, it was not up to the standard of 22 percent on invested capital that was the Sperry-Rand mark of success. Furthermore, it was a one-of-a-kind organization having no organic relationship to the main lines of conglomerate activity in electronics and machines. Therefore the plant was considered more trouble than it was worth to the parent corporation.

To save the 270 jobs at Herkimer, Ladd and his associates had to accomplish the following things: In the first place, at the time of the threatened shutdown there were two local groups competing for an opportunity to buy the plant. The local management group had the manufacturing and marketing expertise but lacked financial backing. The other group had access to financial support but lacked essential knowledge of the business. Ladd served as a mediator between the two groups and managed to convince them to agree upon a single management structure.

Before proceeding locally, Ladd and his associates had to persuade Sperry-Rand to sell the plant to an employee-community group. At first, top management was adamantly opposed to this course, estimating that Sperry-Rand could do better financially and have fewer problems if they simply closed the plant and sold its equipment. Ladd organized the political pressure necessary to change the mind of the corporation's chief executive officer. The climactic meeting brought together that executive with representatives of the employees and community people, and with Representative Donald Mitchell and Senator Jacob Javits.

Still, Sperry-Rand did not make it easy. First, top management demanded a nonrefundable $200,000 deposit to indicate that the employees were serious about their purchase plans. The deposit meant that, if the amount were pledged but the sale was not finally consumated, the employees and local management people would lose the money they had put up from savings, personal loans, and mortgaging their homes.

The leaders of local management and the blue- and white-collar unions (IUE) working with Ladd's group held a plant meeting in the local high school. Within 24 hours they had secured pledges of $193,000, which Sperry-Rand decided was close enough to the amount requested to keep the deal open. (In fact, Ladd had a hard time finding someone in Sperry-Rand willing to accept a check for $193,000. It looked as if the demand for the deposit had been made with the expectation that this would terminate negotiations and now top management found itself embarrassed by the prospect of taking nonrefundable money from employees whose jobs they were terminating.)

Now Sperry-Rand set a time limit of ninety days to raise the $5.7 million estimated as the purchase price. Ladd persuaded three banks in the Mohawk Valley to pledge among them $2 million in loans, on the condition that the Economic Development Administration put up a $2 million loan as a second mortgage, and the employees and community people raise more than $1.5 million in equity money.

Because Sperry-Rand refused to guarantee the figures on the operation of the Herkimer plant, it was impossible to register the stock of the new firm with the Securities and Exchange Commission. Ladd worked out an arrangement with the attorney general of New York State for an issue of stock to be sold exclusively to New York State residents. In order to comply with the requirements of the attorney general's office, the Herkimer group had to submit for approval all publicity material regarding the stock sale, including radio and television commercials. Ladd arranged to have a member of the attorney general's staff work on a day-to-day basis with the campaign to secure quick approval at every necessary step.

In working out the arrangements for the stock sales, Ladd had to secure a feasibility study regarding the new firm. Here he received essential help from the EDA-supported University Technical Assistance Center at the State University of New York at Plattsburgh. He also obtained invaluable free assistance from professionals in banking, finance, accounting, and law.

The planning group rejected major offers for stock purchases by individuals who wanted to make a deal: positions on the board of directors, a long-term lease on a warehouse, and so on. The decision was to organize a broadly based stock selling campaign, bringing in large numbers of small investors. The campaign offered shares at $2 with a minimum purchase of 100 shares and a maximum of 25,000 shares.

Ladd and his associates mapped out the selling campaign as if it were a United Way drive. They divided up the territory and the organizations, assigning individuals to canvass each area or organization.

The canvassing was accompanied by an unprecedented media campaign. Brokerage houses sold the stock without commission. People who could not buy the stock in person were allowed to do so by telephone, and the campaign people sent someone to them to pick up the cash. In a depressed area with an unemployment rate then running well over 11 percent, thousands of people bought stock who had never bought stock before. In fact, some of them were surprised to learn that the stock they purchased might actually yield dividends. Clearly, the motivation was to support the community program rather than to make money. With over $1.7 million raised in stock sales, the total number of stockholders reached 3,500. Employees and local management people purchased 34 percent of the total.

Since the birth of the Mohawk Valley Community Corporation, jobs have been maintained in Herkimer. But that is another story. The purpose of the story told so far is simply to indicate the variety and complexity of the activities necessary to support economic growth and stability in an area.

COUNTY INDUSTRIAL DEVELOPMENT AGENCIES

The Chatauqua County Industrial Development Agency played a major role in the revitalization of Jamestown, New York, and its surrounding area. In a sense, this IDA was invented by David Dawson, a young man who was born and reared in Jamestown. In 1972 he met with county officials to urge them to establish such an organization. Furthermore, he outlined the potential functions of the IDA and presented a detailed job description for the position of administrative director. Impressed by Dawson's drive and vision, county executive Joseph Gerace approved the establishment of the IDA and appointed Dawson to head the organization.

The IDA was beginning operations at the time of the Jamestown plant shutdown crisis, and Dawson worked closely with Mayor Lundine and local bankers in the reorganization and refinancing of five plants.

In 1975 the IDA faced another major industrial crisis, this time in Dunkirk, and Dawson used this impending disaster to add greatly to the resources of the county development program. Allegheny-Ludlum Steel Company had announced its decision to shut down two of the plants in its Bar Products Division, one in Dunkirk, the other in Watervliet in Albany county. About 1,200 jobs were at stake in Dunkirk alone.

Dr. Adolf Lena, chief executive of this division, organized a group of 35 officials to invest in the two plants. In addition to a substantial loan from banks, the new firm needed $10 million from the Economic Development Administration. Through the New York State Job Development Authority, EDA made a grant of $10 million to the IDAs of the two counties ($6 million to Chatauqua, $4 million to Albany). The county IDAs then loaned the $10 million to Al-Tech, the new company. This means that, as long as Al-Tech is able to meet its loan obligations, each IDA has a substantial revolving fund on which to build a greatly expanded economic development program.

In the past six years, the Chatauqua County IDA has raised $40 million in industrial revenue bonds to finance local stabilization and expansion activities, and the program continues to expand. Dawson reports 25 new industrial projects in progress. The IDA has continued to work closely with the Jamestown Area Labor-Management Committee on a number of projects, including the $5.1 million loan to finance the rebuilding and expansion of the old Carborundum plant. Although IDA uses its resources primarily for capital investments, it is also playing an important role in human resources development. The industrywide job-skills training programs organized by JALMC are regularly financed by the IDA.

COMMUNITY DEVELOPMENT CORPORATIONS

The controversial character of community development corporations[1] seems due, in part, to unrealistic expectations that these new and unfamiliar organizations, without highly trained and experienced leadership, and with a small and insecure financial base, would accomplish in a few years what private enterprise and government organizations have so far been unable to accomplish. If we take a more realistic view of the CDC as a new organizational model the leaders and members of which are still learning to build and operate, we can find enough examples of effective action on major community problems to merit further attention and public support. In some cities, CDCs have played major roles in housing rehabilitation and the economic stabilization of neighborhoods. The records so far show less activity in the formation of viable production firms, but this trend may be changing.

Forbes Magazine (March 1978) provides an impressive report on the Kentucky Highland Investment Corporation (KHIC). Funded partially by the Office of Economic Opportunity and private funds, KHIC is a private, nonprofit development corporation with a goal of "local uplift and venture capital investment." In a region in which an average annual family income is below $4,000, KHIC has assisted the development of a tent-making business, a fiberglass kayak business, and other small-scale ventures such as a wooden trophy-base business, a hog feeding operation, and a coal truck body business. So far, KHIC has created 317 new jobs with an investment of only $2.1 million.

While with the now defunct National Center for Productivity and Quality of Working Life, Robert Friedman reported on a growing number of community job-creating projects in fields of public concern but where prospects for profits are not high enough to attract private investors. He cites solar energy, energy conservation, and waste recycling as fields in which CDCs have recently become active.

The Westside Community Development Corporation of San Bernardino, California, has been designing, manufacturing, and installing low-cost solar water heaters and collectors in low- and moderate-income housing. In addition to operating an insulation program, Westside currently has an administrative staff of 30 and employs over 200 minority or low-income trainees.

Hartford, Connecticut, has established a Community Energy Corporation that also combines production of solar heating units with insulation. Bronx Frontier Development Corporation is recycling wastes from a huge vegetable market to produce topsoil for vacant lots in the deteriorated South Bronx community. This CDC has received support from the government, Exxon Corporation, private foundations, business, and community residents. Friedman also notes the establishment of recycling projects in Berkeley, California; Duluth, Minnesota; Somerville, Massachusetts; and Washington, D.C.

As politicians and civic leaders become increasingly aware of the potential of such community self-help projects, we can expect to find growing popularity and local and state governmental support for CDCs. Already Minnesota and Massachusetts have established statewide programs to finance CDCs and CDC-sponsored ventures. Pennsylvania offers tax credits to encourage corporate investment in CDC-sponsored projects. CDC supporting legislation is also being considered in other states.

CONCLUSIONS

We live in an era of public disillusionment with "big government" and "big business." As decision makers in remote conglomerate headquarters shut down local plants, citizens are coming to recognize that they cannot depend on big business to create and maintain employment. They value the great American tradition of local initiative and resourcefulness in solving local problems, yet the problems of economic crises are beyond the capacity of the community to solve without outside help. Increasingly, community leaders are seeking ways to combine local self-help with outside assistance.

Reactions to the federal government tend to be ambivalent. People are fed up with the ineffectiveness and waste of programs directed from the center, with the red tape, delays, and rigidities of giant bureaucracies. Nevertheless, community leaders recognize that, for many vital local projects, they cannot move ahead without enabling legislation and/or federal financial support and technical assistance.

This same ambivalence is increasingly felt in Congress. Representatives and senators are coming to recognize that major problems cannot be solved by having the federal government do more to and for the citizens. The federal government can best serve the people by playing an *enabling* role, stimulating and supporting local community development efforts.

We are still in the exploratory stages of working out this new pattern of relations from federal to local levels. In this process, serious mistakes will inevitably be made. In response to media attacks on waste, inefficiency, and dishonesty, Congress is likely to pressure the Washington bureaucracy to tighten its controls. People in local governments are already complaining that the rapid growth of federal paperwork is forcing them to cut down on their action programs so as to have time to report on progress and plans. Furthermore, administrative regulations designed to reduce waste and inefficiency may seriously limit the freedom of action required by the local-level public-sector entrepreneur.

This points to one of the emerging dilemmas of this new pattern of federal-local collaboration. To solve pressing local problems, local people, citizens and public officials alike, may have to work out creative but also unorthodox programs of action. Can the federal government meet its responsibilities for accountability in the expenditure of public funds without imposing rules and regulations, controls, and reporting procedures that destroy the local initiative and resourcefulness necessary to solve local development problems?

As the federal government supplies enabling legislation, technical assistance, and money, we are finding throughout the country enormous local human resources for community self-help programs. At the same time, we see great disparities in performance from place to place.

At one end of the spectrum, we find an economic development district director who plays a primarily reactive role. If people come to him and his associates for help in securing federal grants of one kind or another, he makes the information available and tries to help prepare the proposal. He does not see beyond the traditional activities of trying to help the various units of government obtain grants for sewers, industrial parks, and so on.

At the other end of the spectrum, we find the public sector entrepreneur. Similar to his counterpart described above, he responds to requests for information and assistance from the field, but he does not merely respond. He has a vision of how to help people combine human and material resources to strengthen the economic base of the community and make the area a more desirable place in which to live and work.

The problem for officials at various levels of government is to devise programs to support the creative public sector entrepreneur and encourage more people in the development planning field to learn to pursue these evolving development strategies. The challenge for universities is to describe systematically the role of the creative public sector entrepreneur at the local and district level and then to devise educational and training programs that will help individuals who have this kind of leadership potential to develop their talents.

NOTE

1. For criticisms of and suggestions for this discussion of CDCs, I am indebted to Ann Waterhouse.

2

PROBLEMS OF INDUSTRIAL
PLANT SHUTDOWNS

David Moberg

JUNE 1979

Len Balluck was discouraged, bitterly discouraged. A few months ago he had hopes that the old Campbell steel mill of Youngstown Sheet and Tube would reopen again under worker-community ownership. He and his fellow workers might have their jobs again.

From his own experience of twenty years in the mill and from the studies made by various experts, he was convinced they could make the mill an economic success and prove that their factory had been scuttled by the exploitative mismanagement of the Lykes conglomerate, not by inevitable forces of the market, or Japanese competitors, or environmental regulations.

Sitting in Isaly's ice cream shop in Struthers, where he regularly shares the news of the community with other steelworkers and local leaders over his morning cup of Sanka, Balluck condemned the recent federal government rejection of the request by the Ecumenical Coalition of the Mahoning Valley for a $27 million grant and guarantees of $245 million in loans to reopen the mill.

"We had high hopes of getting 1,600 jobs back," he said, referring to the first phase of the reopening. "Then we get turned down by Washington. Jimmy Carter will have to bear the burden of that. The people at Isaly's say he hasn't done enough for this valley. Don't even come around here talking to me about Jimmy Carter."

Balluck had organized several hundred of the 4,100 laid-off steelworkers into Steelworkers United for Employment (SUE) after complaining last fall that the Ecumenical Coalition, which has led the fight to reopen the mill, failed to mobilize the steelworkers for the project. It had not been easy. Many former workers were skeptical about the plan and its clerical sponsors. Others were just listless and depressed. Nobody wanted to get his hopes up only to have them crushed again. Balluck knew how precarious many of their lives had been since the sudden September 19, 1977, shutdown. "I can tell you about the drinking, the suicides, the psychiatric wards," he said. "I can tell you all these things."

One friend who was making $24,000 a year as a skilled worker now works as a laborer for $11,000 a year. He is comparatively lucky. While calling for support for SUE, Balluck heard the mother of one young worker explain that her son's benefits had run out the previous December. The pressure of still having no job got to him. He was in a mental hospital. Two people Balluck knew had killed themselves. Nearly one-quarter of those laid off had retired early with reduced benefits.

Few of them are willing to talk about their hardships. They are the sort of people, according to a survey taken by a team from the local university, who find it hard to ask other people for help. They are the sort of people who, despite their anger at the Lykes Corporation and at Jimmy Carter, still blame themselves somewhat for not having a job.

It is tough finding a job in the Mahoning Valley now. Over 9,000 people applied when the local General Motors assembly plant at Lordstown announced it would accept applications. As of late last summer, 80 percent to 90 percent of the laid-off workers were still living in the Youngstown area and only 35 percent to 40 percent of them had found jobs. Some are still receiving benefits, but by now nearly all of the financial cushion has vanished.

That financial aid—unemployment compensation, supplementary unemployment benefits, and Trade Readjustment Assistance—"was a pacifier, welfare," Balluck says. Although it made life almost comfortable for a while, it also undermined the workers' sense of urgency and thus hurt the movement to reopen the mill.

But the shocks keep coming. By the end of this year, the Brier Hill Steelworks, employing 1,100 people, will also close, according to the directors of the LTV Corporation, the new owners. Soon the U.S. Steel Ohio and McDonald Works may also be abandoned, throwing 4,000 more workers on what Balluck calls "the industrial garbage pile."

Across town, in the union hall of Local 1462 of Brier Hill, William Vaughan, a 35-year-old black steelworker who had been working at the mill for fifteen years, talked about what he would do when his job ended. "I want to find a halfway decent job, maybe go to college and get a B.A. degree so I'll have something to fall back on. I know one thing, I'm not gonna get another job like the mill, work fifteen to twenty years and lose my job again. I thought about leaving Youngstown three or four times," he said. "But I've lived here nearly all my life." In theory, Vaughan is supposed to be as mobile as capital, shifting with the opportunities. But like so many workers faced with shutdowns, Vaughan saw Youngstown not only as a place of employment but above all as a home and a community in which to live.

Although his wife's part-time job will help the three kids, Vaughan's impending job loss will hit his family in more than its pocketbook. "My father was just getting to the point where he could do something with his life," Kenny, a top student and athlete in high school, said. "He had some extra money to take trips, pay for college. Now it means I have to get a scholarship. I never thought things like this could happen, that management could say, 'You've got fifteen years in the mill. Now we're shutting it down.'"

THE HIGH COST OF CLOSINGS

More people are discovering what Kenny Vaughan has learned at an early age. Business shutdowns can wreak havoc with the lives of individuals and the well-being of communities. Of course, businesses have failed in the past, or they have shifted from one region to another. Due to inappropriate statistics, it is hard to say definitively whether the frequency of shutdowns has increased or not.

People's level of awareness regarding the consequences of factory and other business "terminations" is changing, however. The Youngstown closing and the fight to keep the mill open, unsuccessful as it now appears to be, have heightened the sense of public urgency and of the possibilities for action. Similarly, there has been a growing interest, stimulated by the work of the Ohio Public Interest Campaign (OPIC), in legislation to provide people with advance notice of shutdowns and to compensate workers and communities for the loss.

Although "runaway shops" and "disinvestment" have been on the lips of activists in northern industrial cities for many years, there is a greater sense of urgency now as the scope of the issue widens. This is partly a result of the steadily worsening impact of conglomerate and major corporate investment decisions on the economic vitality of hundreds of communities. It is also partly a result of a drearier general economic prospect. No longer do people have the faith that new industries will arise to replace those that have closed up shop as the entire economy faces a period of uncertainty and slow growth.

"The broadest, most fundamental starting point is the clear assessment that the postwar boom is really over," says Gar Alperovitz, director of the National Center for Economic Alternatives, which supervised the development of the Youngstown community-worker ownership plan. "Second, no one believes there will be a 'return to normalcy.' Therefore, you can't simply allow short-term dislocation. People begin to say, 'What can we do?' The context has changed."

With that changed context and changing perception comes the possibility of a new political thrust that could radically trasform the U.S. economy. There is a growing awareness of the life-and-death power that capital has over communities and individuals, and that there is no democratic accountability for the exercise of that power.

The emerging new movement demands greater public control over investment decisions, financial capital, and choices of business location. It demands that public needs be considered alongside the private balance sheet. It points in the direction of decentralized planning in the interest of local economic vitality.

"It is very narrow to look at the issue only as plant closings," Alperovitz argues. "The issue is community economic health. It's a much broader question. It's partly plant closings; it's partly new entrances; it's partly expansion. In order to give us a broad enough vision and a strong enough moral posture, the issue is the health of American communities, not just one plant closing. That's also the way people see it."

Yet it is usually a factory closing that jolts people into a new awareness, partly because manufacturing is often the center of community economic life, providing the "export" income that helps stimulate other local businesses.

The community often feels that an implicit contract has been broken. Not only the suddenly jobless workers but other businesses, their employees, local government, and other public institutions have relied on the bigger businesses. They pay dearly for the closings. Then some people realize there was no need for the sudden shutdown. Even if the business died for "natural" reasons—such as the inability to compete profitably—with proper notice, communities could plan for the death's effects.

Often, however, the community loss reflects a decision that is only rational from the viewpoint of a single corporation intent on expanding its profit, size, and power, even if that means unnecessarily destroying factories and communities in the process. Especially with the growth of conglomerates and their special strategies there is an increasingly stark choice: Win greater democratic control of capital or accept greater domination of society by capital.

When a factory dumps waste in a nearby stream, other people pick up the tab: loss of clean water, destruction of fish and wildlife, decline of recreation areas, illness and even death, costs of cleaning up after the corporation. Thanks to the environmental movement, there is a growing recognition in the law and public opinion that the costs of maintaining a healthy environment should be assumed by the firm and not treated as an "externality."

When a factory closes down after years of operation, there are also many costs to the workers, other local businesses and their employees, urban institutions and local government, and taxpayers throughout the state and nation. Taking all those costs into account can lead to a much different view of the economic rationality of a plant closing. Yet those costs are regarded, as environmental costs were in the past, as external to the business balance sheet.

The most immediate costs are borne by the laid-off workers. Because of their unusually high unemployment benefits—in large part a public cost—the workers at Youngstown suffered less than many workers would have. Especially because they were in a highly-paid, unionized industry, their long-term earning prospects for the future are grim.

In his research for the Labor Department, economist Louis Jacobson estimates that workers in industries with low turnover—usually those with high earnings, unions, and predominantly male work forces as well, such as steel and auto workers—suffer most from plant closings.

On the average, workers in such industries will lose the equivalent of about two years' earnings—roughly $30 to $35 thousand—in the first five years after the shutdown. Over their working careers, they will lose 10 percent to 15 percent of what they might have earned. Although older workers may be severely hurt financially because they are forced to retire early, Jacobson finds that workers with three to eight years seniority lose most in the long run because their loss affects more working years.

CONGLOMERATE VERSUS COMMUNITY

When the shutdown occurs in a labor market with high unemployment, or in a small labor market—typical of Youngstown and many other older industrial cities now facing repeated plant closings—the losses are even greater. If unemployment is one-third greater than the national average, the loss can double in a given year. Seeking jobs in a small labor market again boosts the loss. The figures worsen by half again if all the men who drop out of the labor market are included. So steelworkers in a small, depressed community might suffer earnings losses of 30 percent to 50 percent as a result of a shutdown if these effects are compounded, not counting the loss of above average benefits.

After a factory closing, many women typically drop out of the labor force. If we count their loss of earnings along with the loss of women who return to other jobs after a layoff, then women as a group suffer proportionately larger earnings losses than men. Men in high-turnover industries are more likely to be out of work from time to time, to accumulate few seniority benefits and to work in less-skilled jobs than men in lower-turnover industries. These men—in fields such as cotton weaving, television and electronic component manufacture, toys, clothes, and shoemaking—lose proportionately less than men in autos, steel, meatpacking, aerospace, or petroleum refining, Jacobson reports.

Workers can, of course, move at their own expense, often taking a loss on the investment in their home. Because young people are those most likely to move, the community future is hurt also. But family and community ties hold many workers. A study of Youngstown Sheet and Tube workers by Policy Management Associates (PMA) indicated that only one-fifth were thinking of leaving. This finding is not surprising: 77 percent of them had lived in the area for over twenty years and only 16 percent had been born more than 200 miles away.

When a factory shuts down, the effects spread quickly—to suppliers, retail businesses, wholesalers and transportation firms, and various service agencies. The PMA study estimated that an additional 1,650 to 3,600 jobs would be lost in the Youngstown area as a result of the Campbell works shutdowns. Other studies have estimated the indirect job loss at 11,199 to 13,000. Using the PMA estimate, indirect job losses would cause a retail sales drop of $12.2 to $23 million each year, pushing the total sales lost to the range of $66 to $102 million a year.

These costs exact a collective public toll as well. The same PMA study estimated that in the first 3¼ years after the shutdown, local communities around Youngstown would lose up to $7.8 million in taxes, the county would lose $1.1 million, the state up to $8 million, and the federal government up to $15.1 million—a grand total of between $26.8 and $32 million.

At the same time, the cost of the various relief programs—mainly Trade Readjustment Assistance—would run between $34.2 and $37.9 million. By this accounting, the public loss from the shutdown could reach nearly $70 million in slightly over three years.

Even these sums of direct public and private expenses due to a plant closing are inadequate measures. A massive shutdown or a series of smaller closings can disrupt the fabric of the community, upsetting the network of local business transactions and precipitating failures, threatening the quality of public services such as schools, and undermining civic institutions. (Corporate and payroll contributions to the Youngstown United Appeal, for example, dropped by nearly half in the first year after the shutdown.) Especially in a small town, a factory closing can destroy the focus and meaning of community life as a whole.

The most tragic aspect of plant closings, however, is revealed in the stories traded by workers in the Isalys of industrial America—the stories of depression and despair, of broken spirits, and broken marriages. They can be seen, too, in statistics in scientific studies and social work agencies, and even there the figures are understated. As Sidney Cobb and Stanislaw Kasl remarked in their conclusion of a study of physiological and psychological effects in two plant closings, "In the psychological sphere the personal anguish experienced by the men and their families does not seem adequately documented by the statistics of deprivation and change in affective state. . . . The numbers don't seem commensurate with the very real suffering that we observed."

Yet the statistics are bad enough. They found an increased frequency of ulcers in the laid-off men and their wives, greater likelihood of future heart ailments and diabetes, greater hypertension, and more swollen joints. Most of the men compared the experience of the factory closing with the stress of getting married (midway on a scale of life events where 10 equals a traffic ticket, 80 divorce, and 100 death of a spouse), but more than one-fourth found the experience as shattering as divorce or more so. It took most people close to half a year to recover, but as time passed those who were still unemployed tended to blame themselves for their plight. Some became convinced they could not hold a job. Others eventually turned to suicide—at a rate thirty times greater than the national norm.

Similar conclusions can be drawn by projecting the results of a study of the consequences of unemployment. Harvey Brenner, in a study for the Joint Economic Committee, concluded that a 1 percent increase in national unemployment over six years has, in recent decades, been associated with 36,887 deaths, including 20,240 from heart ailments, 920 suicides, 648 homicides, 4,227 state mental hospital admissions, and 3,340 state prison admissions. Counting only the workers in the Youngstown area directly dumped by the Lykes Corporation, Brenner's figures would suggest that the single plant closing will lead to over 130 additional deaths.

Plant shutdowns bring on more family quarrels and violence, mental health problems, and alcoholism. In Youngstown, for example, the Help Hotline had twice as many calls the January following the Campbell shutdown as it had the January before, and the numbers continued to increase. Calls about battered women, child abuse, and family or marital problems tripled in the year following the shutdown. The local Alcoholic Clinic has seen a rise in the number of steelworkers seeking treatment. Referrals to the Eastern Mental Health Clinic doubled in the year after the shutdown.

Adding up all these costs provides one side of what economist David Smith has labeled "the public balance sheet." Benefits of any action taken should be weighed in the same balance. The results are often surprising and quite at odds with the private accounting. For example, Smith assumes that the government should expect a 9 percent return on its money invested in an attempt to save the Youngstown economy. Even the conservative estimates of the PMA study suggest that reopening the Campbell works would bring in enough tax money to justify a $75 million equity investment, far more than the Coalition's proposal for Community Steel would have required. If we add all the other costs and benefits, an even larger public equity investment would be justified.

"What is at issue is the differing perspectives, and therefore cost-benefit calculations, that will be made by an analyst charged with investing on behalf of a public account," Smith writes in *The Public Balance Sheet*, released by the National Center for Economic Alternatives. "Arguments over 'justification' miss the point that the public has a legitimate right to be concerned about the differential imposition of costs and benefits between the public and private sectors."

LIFE AND DEATH POWER

The need for a public reckoning of costs and benefits has never been greater. Communities are now faced with private factory shutdown decisions on an increasingly wide range of "justifications" in the private interest that have less and less to do with the public interest.

As always, many businesses go under that deserve it, although good management could undoubtedly save vast numbers of them. Dun and Bradstreet reports on births, deaths, and moves blame managerial incompetence for 40 percent of business closings. But with the growth of concentrated corporate power, especially in the diversified conglomerate form, and with the expansion of federal intervention in the economy that hastens many business shutdowns, the issue of democratic rights and power is posed more strikingly.

Much of the debate has centered on "runaway shops," businesses that move to the South or overseas in order to pay lower wages, to avoid unions, or to find a highly favorable "business climate."

The shift is dramatic. From 1967 to 1976 the industrial Midwest and Northeast lost 13 percent of its manufacturing work (1.5 million jobs), although the South and Southwest gained 19 percent (900,000 jobs). Also, recent calculations by the economists Robert Frank and Richard Freeman indicate that the rate of direct foreign investment at the beginning of this decade yielded an overall employment loss in the United States of 160,000 jobs a year. (If there had been no overseas investment, they also calculate that U.S. corporate profit would have dropped by roughly 6 percent to 18 percent and that U.S. wages would have increased by roughly 3 percent to 12 percent.)

Overwhelmingly, it is the largest corporations who extend themselves overseas, and, disproportionately, it is also the largest corporations that account for the shifts in capital within the United States. "The larger corporations, using their financial strength, are the first to redistribute their operations out of declining areas into growing ones," writes David Birch, director of the MIT Program on Neighborhood and Regional Change, in *The Job Generation Process*. "They do not hesitate to locate branches in greener pastures, placing an ever greater burden on the smaller firms in struggling areas like the Northeast."

Using data collected by Dun and Bradstreet—because the federal government keeps no useful records on location of firms—Birch argues that the job losses in the North are very rarely the result of an employer picking up and moving the facility south, although that was certainly true in the past with some industries, such as shoes and textiles.

His study also shows that businesses die at about the same rate (5.6 percent to 6.7 percent each year in this decade) in the North and South. But in the faster-growing states, nearly twice as many new firms are born each year and existing firms also expand much more rapidly.

Although the Dun and Bradstreet listings understate the actual migrations, these statistics suggest that the capital shift is often more subtle. Businesses expand and modernize in the South and are gradually allowed to die in the North, with nothing created to replace them. "It is differential branching, not physical migration, that causes many of the regional differences in job growth," Birch writes. "Also, branching seems to be growing in importance over time. Branching is more important in manufacturing than in other sectors of the economy."

Although some economists use Birch's data to argue that runaways are unimportant and that the proper response of the old industrial states would be to make business feel more loved, the statistics do not really erase the fundamental problem: corporate capital's power over the health of local communities. If anything, they highlight the problem. For example, Birch notes that between 1960 and 1976, small firms (under twenty employees) generated 66 percent of all new jobs in the United States. What did the giants, with over 500 employees, do? They generated 13 percent of the total number of jobs created. Even worse, in the Northeast the proportion of new jobs actually decreased by 33 percent in the biggest firms.

The Dun and Bradstreet data may overestimate job creation by small firms, as Hal Wolman of the Urban Institute argues, but the change from the past, when larger firms generated more jobs, is remarkable. What has happened? Birch is not certain, but he suggests that when firms reach a certain size they become multinational and then they "may make all their differential investment overseas."

Why do the big corporations shift their investment? Certainly in many cases it has been to take advantage of cheaper labor—the notorious dollar or two a day for labor in Asia or the low wages offered in the rural South.

A number of researchers point to what may be even more important than cheap labor: greater corporate control. Bob Goodman, author of the book *The Last Entrepreneurs*, argues, for example, that "the North-South shift is in some ways accurate, but it is also very misleading. Rates of growth have been increasing in some northern states, but those are the ones with the strongest anti-labor laws. If you group the anti-labor states, then the absolute number of expansions over the past eight years has been more than double the other states."

Boston University economist Barry Bluestone, who has been studying the New England aircraft industry, argues that some corporate shifts of capital are designed to construct dual lines of production, often including dual subcontractors, in different regions or different countries to avoid disruption by labor.

Others suggest that even anti-union right-to-work laws in the South are not as important to most big businesses in themselves as they are cherished as an indication of a favorable "business climate."

Birch points out as well that New England is no longer a high-wage area and that many of the most rapidly growing Sunbelt cities— Houston, Dallas, Los Angeles, San Diego, and others—are not low-wage areas. Avoiding unionization, according to Birch, is one of the most important reasons for the capital shift, along with factors such as geographical preferences of executives and avoidance of high personal taxes for management (but not corporate tax abatements, which nearly everyone agrees have very little influence on business location decisions).

"Wages are not totally negligible as an influence," he acknowledges. "Going abroad they're quite important." Now, some businesses, having shifted once to the South for low wages, are continuing their shift overseas.

If control is the name of the game—with wages still important in certain circumstances—then the emergence of the conglomerate fits into the picture even more appropriately. The large corporation, and especially the diversified multinational conglomerate, seeks to escape as much as possible from any interference with its control. It hopes to avoid or master competition, business cycles, labor disputes, shifting tastes, national boundaries, and legislation. It cannot, of course, succeed completely, but it can—and does—try.

Partly because the conglomerate has such control, many of its business decisions are made on a basis that might otherwise seem peculiar to a small entrepreneur. These peculiar decisions are extremely important for the issue of community economic health and plant shutdowns. Many of the businesses now being abandoned—and their host communities with them—need not be scrapped by most conventional reckonings, certainly not by any comprehensive public accounting. Only the scramble for conglomerate power and accumulation dooms them.

Youngstown Sheet and Tube is a classic case, as Ohio Public Interest Campaign research director Ed Kelly has demonstrated convincingly. The Lykes conglomerate took over Youngstown Sheet and Tube in 1969, borrowing heavily to buy the much larger steel firm. But rather than use its healthy cash flow to modernize the mills in Youngstown, Lykes made other acquisitions and acquired additional debts as it built its empire. Having failed to modernize the Youngstown works sufficiently, it could not take full advantage of the steel boom in 1973-1974. Then the heavy recession hit. Lykes still owed very large interest payments from its acquisitions. It could not raise the money to modernize at Youngstown, even though its mills there were performing more profitably than the modernized Indiana plant. It decided to abandon Youngstown, later merging it with LTV Corporation. That merger gained the approval of Attorney General Griffin Bell, even though it was anticompetitive and worsened Youngstown's plight by precipitating the closing of Brier Hill and adding new obstacles to the reopening of the Campbell works.

Lykes's behavior was typical of a conglomerate. An acquired firm was raided as a "cashbox" to expand conglomerate control. It had long been true that large corporations, more than local capitalists, felt no attachment to a particular community, but with the conglomerates there is even little attachment to a particular industry; capital in the abstract is everything. The result, however, is frequently very poor management of any one part of the conglomerate.

"People commonly assumed that a big company would not shut down a plant if that plant were making a profit and that, further, if a big company could not operate the local plant at a profit, then the plant was inevitably doomed to failure," Cornell University professor William Foote Whyte wrote in support of a bill aiding worker-community takeovers. "Furthermore, it was assumed that plant shutdowns were a painful but necessary part of the natural process of economic life." But the behavior of conglomerates makes a mockery of those assumptions.

Whyte, for example, found that the Saratoga Knitting Mill began losing money under conglomerate management because the dominant firm's sales force ignored the products of the acquired subsidiary. As an independent unit again, the mill thrived. In another case, Sperry-Rand acquired the Herkimer plant, which made library furniture. Despite Herkimer's long history of profitability, Sperry-Rand closed it because library furniture did not fit into their corporate strategy and because the plant was not making the very high target profit rate—22 percent return on invested capital.

"If Sperry-Rand could make more money elsewhere by shifting its investment out of the Herkimer plant, then the shutdown made good economic sense to the top management of the conglomerate," Whyte wrote. "But it certainly did not make economic sense to the 270 employees, nor did it make sense to the rural people who earned $875,000 a year selling trees to the plant," or to local businesspeople and politicians.

Belden Daniels, a city and regional planning professor at Harvard, argues that "the conglomerate almost invariably imposes costs on the local firm that are diseconomic." For example, the Esmark conglomerate forced its subsidiary, National Tanning and Trade Company, to buy skins for more than the market price. Frequently, small firms gobbled up by a conglomerate are saddled with unneeded overhead and administrative costs that are part of a giant, centralized operation.

One of Daniels's students pointed out another conglomerate tactic that results in irrational plant closings—the calculated tax loss. In the case of National Tanning, "the unfavorable return on the plant was

exaggerated on paper because the parent company apparently manipu-
lated the accounts to produce even greater losses, presumably for tax
shelter purposes. This was achieved by attributing various overhead and
administrative costs incurred" by other plants to the plant that was shut
down.

Conglomerate apologists claim that the takeovers can bring new
managerial skills to the small firms, but often conglomerate control
lessens the needed flexibility of the local unit. Central managers also
frequently lack the specialized knowledge to make good business
decisions for the small unit.

Such mismanagement was a problem in the case of American Safety
Razor, according to Daniels's Harvard study group project. Philip
Morris, which had acquired American Safety Razor, later merged with
Miller Brewing. Safety razors were a tiny part of the new conglomerate.
They no longer fit into the conglomerate marketing strategy. The razor
division was also less profitable, although not unprofitable.

Philip Morris decided to sell it. But the sale was blocked by the
Federal Trade Commission as anticompetitive. So Philip Morris decided
to abandon the firm, even though it would be a hard blow to a
community that had recently suffered four other plant closings.
Eventually, as in the case of Herkimer, National Tanning, and Saratoga
Knitting, there was a management-employee buyout and the firm
prospered with a new marketing strategy under the immediate control of
the local firm.

The American Safety Razor case is an example of a new shutdown
problem. Increasingly, Federal Trade Commission officials say, corporations will threaten to shut down if they are not allowed to merge.
Corporations are continually trying to expand the "failing company"
defense against antitrust charges. If a company is about to go bankrupt,
judges have ruled, a merger can proceed even though it would otherwise
be anticompetitive. LTV Corporation used this argument last year when
it merged with Lykes Corporation, even though Lykes was not failing.

Now that we are in the midst of a new wave of corporate mergers, Ed
Kelly of OPIC predicts that, "based on past experience, we'll see more
plant closings in states like Ohio and Pennsylvania, but also in the
South." There may even be a new rationale for closings. One of the
hottest business consultant strategies of the moment states that
conglomerates should concentrate on dominance of particular markets.
If they cannot dominate, they should close the division, even if it is
profitable.

That would accelerate the irrational closings already caused by
setting arbitrary, high-profit targets. Bennett Harrison, associate pro-

fessor of economics and urban studies at MIT, points to numerous conglomerates that set extreme standards of profitability—such as 25 percent return on investment—or growth and then shut down every branch that could not meet the standards, even though they were several times more profitable than the rest of the industry. "There is nothing especially 'natural' about being unable to do three to four times better than your competition," he argues.

Ultimately the question comes down to who is in control and whose account books count. The concentration of capital, especially in conglomerates, accentuates the conflict between social needs or social rationality and the dictates of private profit-making. Especially at a time when there is insufficient general economic growth to provide balm for the civic wounds, the contradiction between communities—and ultimately the whole nation—and capital finds an acute expression in the problem of plant closings. It is less and less possible to dismiss those shutdowns as representing the triumph of efficiency and the rationality of the market, for they are quite often neither.

Tim Nulty, a former economist for the UAW who worked on plant-closing issues for the Federal Trade Commission, argues that a societywide analysis of "inputs" and "outputs" would reveal that many of the shifts of capital that benefit the private corporation are inefficient. "Does it make sense to take an action with no net increase in national output [as many factory relocations represent] and $100 million cost that is imposed on society? When you net everything out—and that's the definition of efficiency—for many shutdowns there is a real loss of national efficiency."

Nobody makes that national accounting now. Nobody watches the public balance sheet. The conglomerates are accountable to nobody.

SHUTDOWN

Washington, in the end, turned a deaf ear on the request for government assistance in financing the reopening and modernizing of the Campbell mill.

The people in Youngstown who had supported the "Save Our Valley" campaign were furious. The principal reason offered for the rejection was that the Economic Development Administration had gone on record to Congress pledging not to grant loans to the steel industry of over $100 million. But, the coalition replies, White House assistant Jack Watson had gone on record to them in a meeting, a press conference, and a letter last fall that even $300 million was not an outlandish request and was "within the capabilities of the government."

Robert Hall, assistant secretary for economic development in the Commerce Department, argued that the plan was not feasible, but the coalition was dumbfounded that his analysis did not seem to take into account any of the recent developments. For example, in arguing that they had not arranged sufficient equity funding, Hall did not even mention the $10 million that the state of Ohio would put into the project.

There was no acknowledgment that the United Steelworkers, who had been cool, then warm, and then cold toward the plan in the past, had recently come out forcefully for establishing Community Steel, Inc. They had also agreed that all steelworkers hired would start without accumulated seniority because it was a new company. That move alone guaranteed the community-worker plan a 21.4 percent saving in labor cost over earlier estimates.

There had also been a new market study by a well-established consulting firm that demonstrated a strong market within 200 miles of the mill for the full output without any need for special government purchases. Steel industry officials are now admitting there will be a steel shortage by the 1980s, which would further assure the success of and need for Community Steel. However, those same officials—according to a study of the industry by the Argus Corporation—want to cut back all the older U.S. mills so that when the shortage arises, prices will be driven up rapidly. The Argus research indicates the plan is modeled on oil industry patterns: In the tight market, foreign imports will soar in price on the short-term market, providing a cover for U.S. companies to raise their prices.

This is part of the reason the steel companies have fought the "socialistic" Community Steel proposal. Their direct pressure on the Commerce Department and indirect influence through a few traditional consultants was apparently sufficient to kill the plan before it was even completed.

Reverend Charles Rawlings, coordinator of the coalition, says that his search of documents on the case provided through a Freedom of Information Act inquiry showed no sign that the new proposals were ever even read. One government development economist speaking off the record confirmed Rawlings's fears: The final proposal was never considered. A former skeptic about the viability of the original plan, this economist was now convinced that the revised version could have worked.

There is only one long-shot hope left for Community Steel. If Carter wants to be reelected, he needs the support of the citizens of Ohio, and for that he needs to garner support in the Mahoning Valley. He cannot obtain it without doing something dramatic, such as funding Community Steel. There are other proposals—such as putting a giant central coke oven or a sponge iron facility into operation—but they provide few jobs and have other drawbacks.

Although many members of the coalition are ready to throw in the towel on Community Steel, the movement started there is still developing. A Tri-State Commission on the steel industry involving labor, church, and community groups has been formed. The Commission has filed objections to U.S. Steel's application with the Army Corps of Engineers to build a new-from-scratch "greenfield" steel mill on Lake Erie in Conneaut, Ohio. They demand that alternative sites, such as Youngstown and Pittsburgh, be considered. The new strategy emerging from the steel communities emphasizes "brownfield" development, rebuilding the steel industry in the communities where steelworkers live.

The local union at Brier Hill waged a spunky fight against the LTV decision to close the mill, offering counterproposals for reopening it under worker-community ownership and carrying their protests into the local country club meeting of steel executives. They battled an apparent plan to shut the mill this spring, but then agreed—mistakenly, many feel—to cooperate with an orderly shutdown of the mill later this year. They continue to press for alternatives and have helped inspire the recent complete turnover of local union leadership—except for their own local—that may prepare the Youngstown labor movement for a stronger role in any future contest over closings of U.S. Steel.

"The community effort here was the best effort ever made," said John Barbero, retiring vice-president of Local 1462. "But I don't know where we're going now. I'm very pessimistic. A good part of the problem was that people were just not getting involved. The effort was never really made on the people to organize them." Despite the impressive coalition effort, there was always the undercurrent of discouragement: Why didn't people—especially the affected steelworkers—back the plan more forcefully?

Many point to the extensive benefits as having "bought off" the workers. Others suggest many people felt the project was impossible and found it hard to believe that a group of clergy knew anything about steel. The on-again, off-again lukewarm support from the district and international steel union officers hurt. The coalition had broad support: 80 percent of the area residents showed a positive reaction to the Coalition in a poll last fall, compared with 18 percent showing a positive response to Carter. But it was shallow support, coalition attorney Staughton Lynd says.

There are other, deeper cultural problems that can only be overcome as a movement convinces people of their capacity to act and of their right to make demands. In that sense, the task is similar to starting a labor movement or a civil rights movement. "The thing that people say so often: 'But it's *their* property,' " district union representative Marvin Weinstock said. "People don't think they can affect it, something so big. It's been pounded away that they have no control when it's someone else's property. People do not yet feel that their rights are on a par with—or superior to—the rights of property, even when they have been deeply hurt."

Likewise, people have so little experience in democracy and direct control of their lives and often have so little knowledge of the industries on which they depend, they feel they have no capacity to act. Many of the steelworkers from Campbell, however, were anxious to use their skills to open the mill. They knew how to run the mill better, how to save money, how to work together in a way the old management had hindered with its authoritarian rule of the workplace—if only somebody could obtain the money to start the mill rolling again.

Top-down control in the labor movement denies workers the one major opportunity they have for exercise of democracy and building a sense of their capacity for self-management, too. "When the union took away our most powerful weapon—the strike—when they wouldn't give us the right to ratify our contract democratically, when we were told for years not to rock the boat, and then when all these pacifiers came in—that's why nobody took action," Len Balluck says.

Why was there so little action from other workers? "Suppose you're my neighbor," Balluck explains. "You have a good job. You don't give a damn. Too bad, but as long as money's coming into my pocket, I don't care. That's what it's come to. People don't care as long as their pockets are full."

It may take good ideas, solid plans, technical expertise, access to money, and sufficient clout to elect sympathetic politicians or to force other legislators to respond in order to turn the tide against conglomerate shutdowns, to assert the primacy of the public balance sheet, and to defend economic health of communities. Above all, however, it takes a dramatic cultural shift in favor of democratic initiative against the power of capital. That requires a powerful political movement.

POSTSCRIPT: MARCH 1985

The powerful political movement did not materialize in Youngstown. Nor has it flourished elsewhere in the nation. The Reagan promises of growth through unrestrained capitalism won the day, especially when the Democrats had no persuasive alternative strategy for growth and employment. In response to the cry from the Youngstowns of the Northeast and Midwest, some Democratic political leaders talked about development of industrial policies, but the focus of such proposals quickly shifted from jobs—especially jobs where the jobless were—to international competitiveness. Even modest steps for government intervention, such as a federal investment institution, were dropped, and by early 1985 industrial policy itself seemed to have become yesterday's fashion.

The rise of neoliberalism within the party could be exemplified by James Fallows's article in the *Atlantic* of March 1985. In the misleading guise of advocating a policy oriented toward "people" rather than "places," he offered essentially a very slightly guilty version of the conservative admonition to the victims of plant closings: Go somewhere else. Fallows at least thought the government should provide shoes for the hike, but insulted anyone who would not flee his or her community as probably uncreative and not socially worthwhile anyway. But throughout the country, many people are more aware that the welfare of people is not so easily divorced from the welfare of the communities in which they live. Usually implicitly, sometimes explicitly, they realize that the real issue in any case is the social control of investment and the accountability of private business to the public.

If America was back, as Reagan maintained, it left Youngstown behind. By 1984 all the basic steel manufacturing in Youngstown was gone. One small minimill and related fabrication facilities survived at the old U.S. Steel McDonald Works. A pipe mill barely limped along at part of the old Campbell Works; another pipe mill had gone bankrupt. Steel mills in nearby Warren were also on the ropes. After steel closed, there were waves of closings of related manufacturing companies, then retail and service establishments.

The deep depression of 1981-1982 cut employment in half at the previously thriving General Motors plant in nearby Lordstown, and the militant local there even made some local concessions. GM shifted much of the work at its huge Packard Electric factory in Warren overseas, cutting employment nearly in half until the local agreed to a two-tier wage system. When 300 new reduced-wage jobs were announced in 1985, 30,000 people applied. In 1984 and early 1985 around 17 percent of the local work force was unemployed, but as Youngstown State University labor studies director John Russo estimated, if people involuntarily working part time or retired were counted, the real unemployment rate would be closer to 33 percent. Despite the deep ties to the community reported at the beginning of Youngstown's travail, an exodus had drained thousands who often traveled south and west only to find quite often more unemployment, poverty, and hardship in an alien community.

The fight against plant closings did escalate after the initial Campbell Works battle. When Lykes Corporation, owner of Youngstown Sheet and Tube, merged with Ling Temco Vought, owner of Jones & Laughlin Steel, the Brier Hill works at Youngstown was the victim. The local union fought the legally questionable merger, but it was foiled by opposition from its own international union leadership and by lack of local support.

The struggle then shifted to U.S. Steel, where local management had promised to keep the mills open as long as they were profitable. But in late 1979 it announced the mills would be closed. As at Brier Hill, local union leaders—assisted by a few veterans of the earlier fights, such as attorney Staughton Lynd—took the initiative, not the religious and community leaders who had sparked the Campbell worker ownership drive. In an important lawsuit, Lynd argued that steelworkers had "detrimentally relied" on the promises of U.S. Steel to keep the mills open. Meanwhile, steelworkers rallied at the local U.S. Steel administration building, then marched down the hill, broke open the door, and occupied their building with the message, If U.S. Steel won't run these factories, let us workers buy them and operate them. Much to their later regret, the sit-in protestors agreed to leave after promises that company officials would meet with them.

Lynd was able to persuade Federal District Judge Thomas Lambros to enjoin U.S. Steel from closing the plant. In his judgment, Lambros outlined a provocative concept of community property rights. In part, he said, "It seems to me that a property right has arisen from this lengthy, long-established relationship between United States Steel, the steel industry as an institution, the community in Youngstown, the people in Mahoning County and the Mahoning Valley in having given and devoted their lives to this industry. Perhaps not a property right to the extent that can be remedied by compelling U.S. Steel to remain in Youngstown. I think the law could not possibly recognize that type of obligation. But I think the law can recognize the property right to the extent that U.S. Steel cannot leave Mahoning Valley and the Youngstown area in a state of waste, that it cannot completely abandon its obligation to that community, because certain vested rights have arisen out of this long relationship and institution."

Lynd made a compelling case that the Youngstown mills had more than covered their own fixed expenses, which the local manager had understood was the criterion of profitability. But Lambros denied there was a contract or that the plant was profitable as he dismissed the case.

Local steelworkers and community supporters had drawn up a plan for Community Steel, which would have used $23 million in federal funds from the Economic Development Administration and an Urban Development Action Grant to set up an employee stock ownership plan (ESOP). With $100 million in federally guaranteed loans and another $11 million in unguaranteed loans, the new firm could employ nearly 750 people and make a profit within five years—far from the original 3,600 workers, but still significant. Consultants concluded that many profitable, specialized products from U.S. Steel's former lines could continue to be produced, but that U.S. Steel was not interested in such small-volume business. U.S. Steel vehemently resisted any discussion of sale to such a "socialist" enterprise, and Lynd, on behalf of Community Steel, brought an antitrust action. But U.S. Steel cut the ground out from under them by arranging to lease a small portion of the mill to two local businessmen.

As Lynd reflected on the struggles in his intriguing book, *The Fight Against Shutdowns* (1982), he downplayed the value of his own legal efforts and emphasized the effectiveness even in such an unlikely situation of strikes, sit-ins, demonstrations, and other direct action. He also noted that plant shutdown protests must strike quickly; demoralization sets in quickly and a movement is hard to sustain for many months. Although he argued persuasively that investment in traditional steelmaking regions made economic and social sense, the deep depression of 1981-1982, which continued long afterward for steel, made little investment attractive. Federal funds would be necessary, especially to make community ownership viable, he believed. Yet it is also true that federal policies—such as a commitment to rebuild railroads, repair bridges, or fix crumbling urban roads and sewers—would be equally critical in maintaining demand for steel.

Although the Youngstown movements died with little success, the examples from the battles there inspired people elsewhere. Worker buy-outs, through cooperatives or ESOPs, have become more common in plant closings. Many unions, including even the steelworkers, have now begun to give more support to such projects, although there is still deep suspicion of them among labor leaders.

In the immediate area, there are several progeny of the Youngstown movement. When National Steel announced in 1982 it would shut down its Weirton Steel mill in the West Virginia town of the same name, the local independent steelworkers union actively supported the new ESOP. But a core of workers believed that the company-union plan demanded unnecessarily large financial concessions and gave workers too little control over "their" factory. Yet as long as the new Weirton ESOP relied on large private loans, bankers would insist that management and outside directors be in control. Under that pressure, plans for the ESOP's structure grew less democratic as discussions continued.

In the Pittsburgh area, the Tri-State Commission on Steel—covering Pennsylvania, Ohio, and West Virginia—sought to confront the widespread closing of steel mills by U.S. Steel and others with a new plan for a Steel Valley Authority. Partly modeled on the Tennessee Valley Authority and on Conrail, this quasi-public authority would have the power of eminent domain to acquire steelmaking properties and assemble a coherent, viable industry in the region out of the fragmented pieces left behind. The idea of using powers of eminent domain to fight plant closings has already been invoked in other conflicts. The mayor of New Bedford, Massachusetts, effectively threatened Gulf & Western, owner of Morse Tool, to sell to another owner rather than close the factory.

In contrast to Weirton, the Steel Valley Authority plans for community and worker control, government grants and guaranteed loans, and use of eminent domain. In February 1985, the Tri-State Commission received a consultant study showing how much of U.S. Steel's Duquesne Works could be saved and made profitable under either alternative private or public ownership. So far the mill has been saved from the wrecking ball and from U.S. Steel's plans to use the land for real estate projects—a manifestation of its moves to get out of the steel business and into other, more profitable fields. (U.S. Steel, after its acquisition of Marathon Oil with the tax breaks that were supposed to finance reconstruction of the steel industry, is now listed in the Fortune 500 as an oil company.)

Some steelworkers, community leaders, and clergy have adopted a controversial attack on local banks, corporate officials, churches, and other institutions as complicit in the "corporate evil" that is resulting in disinvestment from the Monongahela Valley steel industry. Although not all opponents of the steel companies approve of their tactics, such as disrupting church services or putting fish in bank safe deposit boxes, their protests have highlighted the problems and made the Tri-State Commission seem like a more moderate alternative.

Despite the hostile national political climate, a movement for greater worker and community control of businesses and local economies has grown, winning occasional victories and fighting many losing battles throughout the country. Youngstown may not have seen the benefits of the struggles that started there, but other communities and other workers may.

REFERENCE

LYND, S. (1982) The Fight Against Shutdowns. San Pedro, CA: Singlejack.

3

OPTIONS FOR REINDUSTRIALIZATION: BROWNFIELD VERSUS GREENFIELD APPROACHES

Staughton Lynd

Each autumn from 1977 to 1979 in Youngstown, Ohio, a major steel mill announced its intention to close.

At 8 a.m. on a Monday in September 1977, the presidents of local unions representing production and maintenance workers of Youngstown Steel and Tube in the Mahoning Valley received phone calls. They were asked to come to the company's offices at 10 a.m. On arrival they were handed a statement simultaneously being released to the media. It said that more than 4,000 workers at Sheet and Tube's Campbell Works were to be permanently terminated. The layoffs began by Friday of that same week.

In November 1977, the Lykes Corporation, the conglomerate that owned Youngstown Sheet and Tube, and the Ling Temco Vought Corporation, the conglomerate that owned Jones and Laughlin Steel, announced plans for a merger. The merger had to be approved by the

U.S. Department of Justice. The United Steelworkers of America and the Ecumenical Coalition of the Mahoning Valley formally requested that if the merger of Lykes and LTV were approved, the following conditions should be attached:

(1) The merged corporation would commit itself not to close permanently any substantial producing unit of any of its plants without first obtaining permission from the Justice Department.

(2) The Justice Department would make it clear that permission for such closure of a substantial producing unit would be given only upon proof shown by the merged corporation that (a) severe losses would be experienced if the facility were not closed, and (b) there was no way to obtain funds for modernizaton of the facility (including community or government loans or loans guaranteed by the government).

In June 1978, the Department of Justice approved the merger without attaching conditions. That same month Local 1462 of the Steelworkers, representing production and maintenance workers at the Brier Hill Works of Youngstown Sheet and Tube, requested to meet with J and L officers to discuss the fate of Brier Hill after the merger. Workers at Brier Hill were concerned because the product they made there, a component for seamless pipe, was manufactured more efficiently at a J and L plant in Aliquippa, Pennsylvania, fewer than seventy-five miles away. The workers feared that the merged steel companies would require only one of the two mills and that Brier Hill would be closed. J and L officers refused to meet with them at that time. In October 1978, the merging companies sent a joint prospectus to their stockholders stating that Brier Hill would indeed be closed after the merger. That was how the Brier HIll workers received the news.

A little more than a year later, in October 1979, local union officers at U.S. Steel's Ohio Works in Youngstown became concerned about rumors that their mill might be closed. A forum was held in Cleveland on November 1, 1979, attended by several local union officers from Youngstown and by Frederick Foote, a public relations representative for U.S. Steel. Bob Vasquez, president of Local 1330, the Ohio Works local, asked Mr. Foote whether the Ohio Works was to be closed. Mr. Foote answered, as he was quoted in the paper the next day: "We have said all along that the plant has been profitable, and there are no plans for a shutdown" (Local 1330, 1979: 57-58; Warren Tribune Chronicle, November 2, 1979).

Just 26 days later, U.S. Steel announced that the Ohio Works and its sister mill in Youngstown, the McDonald Works, would be closed by June 30, 1980. There was no advance notice to the workers. Upon hearing the news they immediately sought to arrange a meeting with national U.S. Steel management in Pittsburgh. Finally, on December 20, a meeting took place between two representatives of U.S. Steel labor relations and the presidents of the two production and maintenance locals at the Ohio and McDonald works. The workers offered to give up $500,000 a month, or $6 million a year in incentive pay if the company would reconsider its decision. A week later they received a phone call in which they were told, in effect, that the company did not believe further discussion would be fruitful, because the decision to close the mills was irrevocable.

I have a friend and neighbor in Youngstown who worked in the open hearth at the Brier Hill Works for twenty-odd years until it closed in December 1979. He was the vice-president of the Steelworkers' local union there.

One day in late August or early September 1977, John and I were talking, and he asked me if I'd seen a story in the paper about a speech given by Stewart Udall, the former Secretary of the Interior. John said that Udall had thrown out the concept that it was preferable for industrial modernization to take place in what he called a "brownfield," as opposed to a "greenfield."

In a "greenfield" industrialization, a company goes into a hitherto rural area and builds not only a new industrial facility, but also the surrounding social community—the roads, the sewers, the schools: all that is necessary for people to live as well as work. An example of greenfield development is the huge new steel mill proposed by U.S. Steel in Conneaut, Ohio. In the "brownfield" model, one goes into an area (if we are thinking of steel) such as Pittsburgh, Youngstown, or Lackawanna, and seeks a way to rebuild the industry without disrupting the community that exists there.

And John said, "Staughton, this is what we ought to be thinking about the steel crisis. A brownfield rather than a greenfield model is preferable, if possible."

Under cover of arguments about the obsolescence and noncompetitiveness of particular mills and particular locations, the American steel industry is giving up the steel business altogether. The United States is losing the ability to supply its steel needs because steel companies are investing outside the steel industry. Investment in steel is profitable; indeed it appears that the American steel industry may be the most profitable in the world.[1] But investment in steel is not *as profitable* as investment in, say, the chemical industry or downtown realty, and therefore U.S. Steel and other steel companies have been putting their new investment dollars elsewhere.

In the case of U.S Steel, 37 percent of its investment in the years 1975 to 1979 was in expansion and growth of nonsteel businesses (U.S. Steel, 1979a: 9). In the latter four of these five years, the company's nonsteel assets grew 80 percent to $4.7 billion, while steel assets increased only 13 percent to $5.9 billion and steel-making capacity actually decreased (Business Week, 1979; U.S. Office of Technology Assessment, 1980: 80). In 1979, the same year in which the shutdowns of the Youngstown Works and other steel facilities were announced, the company opened a new joint-venture shopping center near Pittsburgh, containing the largest enclosed mall in Pennsylvania and, a few weeks after the Youngstown shutdown announcement, signed a letter of intent with Tenneco Chemicals, Inc., to build worldscale chemical facilities in Houston (U.S. Steel, 1979a: 11). Of the steel facilities at which closings were announced in 1979, at least the New Haven wire mill appears to have consistently turned a modest profit of about $500,000 a year (New Haven Advocate, 1979). It fell victim to the philosophy reaffirmed by David Roderick, chairman of the board of U.S. Steel, in February 1981, that "new spending will go to those businesses that provide the highest rate of return" (New York Times, 1981).

Publicly, Mr. Roderick (1979) bemoans the possibility that the United States may become dependent on steel imports as it has become dependent on foreign oil. Yet in the meantime, the company continues to cut back its steel capacity. There is at least the possibility that the industry, led by its largest company, is deliberately *restricting* steel output so as to be able to charge higher prices. Certain financial analysts recommended this strategy just before the wave of shutdowns began in 1977. Charles Bradford (1977: 26), steel analyst for Merrill Lynch, advised: "The announced expansion plans of the United States steel industry do not make any sense to us unless an equal amount of antiquated facilities are closed." The Argus Research Corporation (1977: 2) of New York City was more blunt:

> By contracting their capacity base, American steel producers will concede a still larger share of the U.S. market to foreign suppliers, but along with this will go increased power to set pricing patterns. This is not unlike the situation that developed in the domestic oil industry earlier in this decade, after which petroleum prices soon began to rise sharply. We expect the same pattern to occur in steel prices.

Tax incentives for investment have been offered by some as a potential solution to steel industry problems. However, there is absolutely no assurance that the kind now under consideration in Congress for industry in general [2] will result in steel industry modernization. Steel companies may take such windfall dollars and invest them outside steel. Joel Hirschhorn (1980), project director for the Office of Technology Assessment steel study, comments:

> Federal policies toward the steel industry benefit the large integrated companies. Nevertheless, these producers are likely to continue to diversify and get out of steelmaking. . . . Measures such as refundable tax credits may only hasten non-steel investments by large steelmakers who have decided to diversify.

As a piece of social planning, the pending tax legislation in Congress is like throwing paint at a wall and hoping for a picture.

The steel industry defends its practice of facility and community abandonment as follows: Plant closings, we are told, are unpleasant but necessary, just like surgery. American industry must be modernized to compete with European and Japanese imports. The way to modernize, according to industry spokespersons, is to close old facilities and build from the ground up in new locations. And government must do its part by removing restrictions on steel industry price increases, relaxing overzealous programs to clean up the environment, and reforming tax laws to make more capital available.

Beneath the surface of the industry's analysis is the long-standing conventional wisdom that capital should have unrestricted mobility. Only if businesses are free to shut down, and free to move elsewhere, it is argued, will entrepreneurs make investment decisions most calculated to keep industries like auto and steel competitive in the world economy. In this view, the factors of production should be located wherever they will yield maximum profit.

Many voices urge this view. Mayor Richard Caliguiri of Pittsburgh concedes that Pittsburgh has no concrete plans for retraining laid-off steelworkers or teaching skills to unemployed blacks. He suggests, in fact, that it may be better for the city's disenfranchised to go elsewhere: "I'd rather have less people with high incomes than more people with relatively low earning and spending power" (Pittsburgh Press, November 7, 1980).

This argument was echoed by the company attorney for U.S. Steel at the close of the trial in Youngstown. He claimed that workers who had lost their jobs could transfer to other U.S. Steel plants if they desired, adding:

> They don't know what . . . being out in the street really means, not like some lawyers do. They are not out of jobs. They only have an inconvenience of moving.

> Millions and millions of Americans every year move for better jobs and move from one city to another city but these Plaintiffs insist they have a contractual right not to move [Local 1330, 1979: 939].

Confronted with this logic, Youngstown workers and their advocates struggled for words to express another viewpoint. Ed Mann said in meeting after meeting, "We're not gypsies." John Barbero recalled how Aneurin Bevan of the British Labor Party had described the uprooting of his family from Wales and had asked, "When do we stop running?"

From the meetings, the kitchen-table arguments, the leaflet writing, and the lawsuits, there has emerged a more fully articulated argument for brownfield modernization. It makes the following points:

(1) Even from the standpoint of the single firm, greenfield modernization is more expensive than brownfield modernization.

(2) When costs to the community as well as costs to the firm are considered, the case for brownfield modernization becomes overwhelming.

(3) In the last analysis, the case for brownfield modernization rests on values that cannot be measured, and expresses an ethical and political choice for a society in which "an injury to one is an injury to all." There is no economic invisible hand that makes the reindustrializing of America in new locations rather than older ones necessary or desirable. Next to anti-unionism, the strongest motivation for flight from existing industrial cities would appear to be the American penchant to abandon last year's car, last year's spouse, and last year's community. The concern for family and community so much talked about nowadays should express itself in a contrary presumption that modernization in existing sites is socially preferable.

Two comprehensive federal studies conducted in the past five years have reached the conclusion that it is cheaper for the individual steel company to modernize in existing locations than in new, greenfield sites.

In October 1977, the Council on Wage and Price Stability concluded that "replacement of existing plants by efficient, new greenfield operations is simply uneconomic at today's capital costs." What the Council termed "rounding out"—that is, adding some new facilities to existing plants—is a more profitable strategy of expansion. Although greenfield expansion results in steel production at an *operating cost* somewhat lower than that of brownfield ("round-out") expansion, brownfield's vastly lower *capital cost* makes the difference: It is approximately $60 cheaper per ton of finished steel. The Council's results were presented as shown in Table 3.1.

A report by the federal Office of Technology Assessment released in the spring of 1980 came to similar conclusions. The report states:

It is accepted that greenfield expansion provides the greatest opportunities for installing optimum new technology and plant layout and offers maximum production cost savings. These advantages, however, usually will not offset the large capital costs. . . . There is agreement that

TABLE 3.1

	Average Existing Carbon Steel Plant	Rounding Out of Existing Plant	New Greenfield Carbon Steel Plant
Operating costs	300	260	240
Additional capital charges (including equity returns)	–	100	177
Total additional costs per ton	–	60	117

NOTE: All figures in 1976 U.S. $/net ton (U.S. Council on Wage and Price Stability, 1977: 82).

greenfield expansion cannot be justified, either on the basis of the price necessary to obtain an acceptable level of profitability or in terms of the *net* increase in costs. The case of energy conservation exemplifies this conclusion: by spending $11/tonne on retrofit equipment, a steel company could save 1.1 million Btu/tonne; a greenfield replacement of the same productive facilities could save 8 times that much energy, but it would cost at least 120 times as much to accomplish. Given current policies and price levels, the capital and financial costs are too high relative to the benefits from the best available integrated steelmaking technology to favor greenfield expansion [U.S. Office of Technology Assessment, 1980: 312].[3]

This conclusion is supported with convergent data from government, university, industry, and consulting-firm studies. Table 3.2 is from the OTA report.

The comparison can be made even more concrete by considering the Ohio Works at Youngstown. William Kirwan, superintendent of the mill, proposed a plan to his corporate superiors for modernization of the Ohio Works by building electric furnaces and a continuous caster. The estimated cost was $208 million. The annual production was estimated at 700,000 to 800,000 tons of steel, so that the cost per ton of modernization would be $350 to $400 per ton. By contrast, U.S. Steel proposes to build a greenfield mill at Conneaut that would produce about 4 million tons of steel and cost about $4 billion to build, resulting in a cost per ton of modernization of about $1,000 per ton.[4] In Mr. Kirwan's words, his plan recommended that "a greenfield plant be built on a brownfield site" that would cost "one helluva lot less dollars than a Conneaut" and be "a far more desirable short and long range alternative to the tremendous cost and the socio-economic impact involved in phasing out the plant."

There is no reason to suppose that the comparative figures for the cost of brownfield and greenfield modernization will change significant-

TABLE 3.2
Integrated Carbon Steel Plant Capital Cost Estimates
for New Shipments Capacity

	Year	1978 Dollar/Tonne of Capacity Roundout	Greenfield
A. D. Little	1975	628	1,296
Fordham	1975	880	1,474
COWPS	1976	710	1,502
U.S. Steel	1976	NA	1,220
Marcus	1976	630	1,514
Inland Steel	1977	520	956
Mueller	1978	715	1,210
Republic Steel	1979	372-636	1,367-1,317
American Iron & Steel Institute	1980	743	1,287

SOURCE: U.S. Office of Technology Assessment (1980: 315).

ly in the foreseeable future. They have been relatively constant for the past quarter century. For example, in 1958, Bethlehem Steel estimated that an entirely new fully integrated plant in the Chicago area of 2.5 million tons ingot capacity would cost $300 per ton ingot capacity as compared to $135 per ton ingot capacity for expansion of Youngstown Sheet and Tube's existing plant in the area (United States v. Bethlehem Steel Corporation, 1958).[5]

A comparison of the costs to the company of greenfield modernization and brownfield modernization is only the first step in an adequate analysis. One must also consider social costs. Even if greenfield modernization were cheaper for the company, it might be more expensive for society as a whole.

Late in 1978 an analysis was conducted of socioeconomic effects of the Campbell Works shutdown. It found that in addition to the 4,100 employees at the works whose jobs were terminated, at least another 3,600 jobs would be lost through the secondary multiplier effect on suppliers, retail businesses, and others. Loss of wages to the former Campbell Works employees during the first three years after the shutdown was estimated at $50 to $70 million, and loss of wages to those in other businesses during the same period at $63.5 million. The study projected costs to the public sector during the same three years of $60 to $70 million. About half of these public costs were expected to take the form of local, county, state, and federal tax losses: In the city of Campbell, for example, the Campbell Works provided approximately 65 percent of the city's property-tax revenues. The other $35 million in projected public costs was expected to derive from various benefit programs, particularly the federal Trade Readjustment Assistance Act,

TABLE 3.3[a]

	British Steel Corporation	Other Nationalized Industries	Total
Annual savings to industry	231	77	308
Annual cost to industry in unemployment benefits, etc.	57	19	76
Net annual savings to industry	174	58	232
Tax loss to national government	–	–	408
Additional welfare payments	–	–	134
Total lost to Exchequer	–	–	542

SOURCE: Iron and Steel Trades Confederation (1980: 76-79). This study cites Rowthorn and Ward (1979). The Iron and Steel Trades Confederation is the leading trade union in the British steel industry.
a. Figures are in millions of British pounds.

which provides benefits to workers who lose their jobs because of imports (Policy and Management Associates, Inc., 1978).

By January 1981, unemployment in the Youngstown-Warren metropolitan statistical area had reached 15.4 percent, the highest level since the Depression (Youngstown Vindicator, March 5, 1981).[6] City after city in the Mahoning Valley has experienced a budgetary crisis followed by wage cuts and layoffs for public employees, and cutbacks in social services. For example, in May 1980 nearly all of Youngstown's 1,000 municipal workers, including firefighters and police officers, went on strike for pay increases the city said it could not provide because of revenue lost in the shutdown of the Valley's steel mills (New York Times, May 3, 1980).

In Great Britain, because the steel industry is largely owned by the national government, it is possible to compare the savings to the government as entrepreneur from mill shutdowns with the costs to the government as provider of social services that the shutdowns entail. The Department of Applied Economics and Faculty of Economics at Cambridge University made such a calculation of the costs and benefits over five years of a shutdown program undertaken by the British Steel Corporation in December 1979. The calculation was as shown in Table 3.3.

As the table makes clear, the estimated effect of the shutdowns on the national government considered both as entrepreneur *and* as provider of social services would be £542 minus the savings of £232, or £310 million (about $600 million) *lost per year*. There is no reason to believe that a calculation of benefits and costs would be significantly different in an American shutdown setting.

To fully comprehend the comparative social costs of brownfield and greenfield modernization, one must also consider the increased social cost of modernization at the greenfield site. For example, Conneaut has a present population of approximately 15,000. Although U.S. Steel estimates that construction of the proposed mill would cause the population to double, others have estimated that the increase in population might be close to 60,000. Furthermore, U.S. Steel has assured local officials that the coming of the mill might make it possible to eliminate property taxes entirely (U.S. Steel, 1979b). But James Williams, a partner in Philadelphia's Murphy-Williams Urban Planning and Housing Consultants, estimates that the development of the mill could cost each resident of Erie, Crawford, and Ashtabula counties $6,500 a year over a period of 25 years. This is his estimate of what it would cost the community to develop services such as sewer plants, water plants, fire and police protection, and school operations, including busing, roads, government administration, utilities, libraries, health care, and recreation.[7]

Felix Rohatyn, the financier who engineered New York City's "survival," has said essentially the same thing. Rohatyn's (1981: 16) words directly confront the investment strategy of the American steel corporations:

> The currently fashionable notion of backing the winners instead of losers is as facile as it is shallow. The losers today are automotive, steel, glass, rubber, and other basic industries. That this nation can continue to function while writing off such industries to foreign competition strikes me as nonsense. . . . We cannot become a nation of short-order cooks and saleswomen, Xerox machine operators and messenger boys.

Still more specifically, Rohatyn (1981: 20) echoes the analysis developed by Youngstown steelworkers:

> In a world where capital will be in shorter supply than energy, is it really a valid use of resources to have to build anew in the Sun Belt the existing schoolhouses, firehouses, transit systems, etc., of the North for the benefit of the new immigrants in the South, instead of maintaining and improving what we already have in place here? Is it rational to think that northern cities teeming with the unemployed and unemployable will not be permanent wards of the federal government at vast financial and social cost? . . . Doesn't the notion of "taking the people to the jobs" completely ignore that many of those people, in large parts of this country, are unwilling and unable to move?

Even the inclusion of social costs in an analysis of the greenfield and brownfield alternatives is not enough. Cost-benefit analysis does not wholly address the issues in the debate over steel modernization, no matter what costs are considered. Some things simply cannot be quantified. The challenge to advocates of brownfield modernization is to find a precise way to talk about values that cannot be measured.

To do this, it is necessary to step outside the political philosophy that derives social decisions from individual rights, and considers merely as another cost, albeit a "social" cost, the destruction of "the complex nexus of family, neighborhood, religion and work that has provided the framework in which most people live out their lives—our communities" (Lynch, 1981).

The effects of investment decisions are similar to impacts on the natural environment: One must consider the *ecological* effect of actions, not merely the aggregate of the action's effects on separate individuals, or on separate aspects of the environment, such as air, water, wildlife, and the like. This is a case in which the whole is more than the sum of the parts. The impact of a shutdown on a community cannot be found by treating separately the effect of the shutdown on employment, crime, alcoholism, the divorce rate, and suicide. The effect of a shutdown on a family cannot adequately be assessed by calculating the effect on each family member separately, and deriving a resulting compound trajectory of family behavior.

Further, although we in this country have an old and useful tradition of discourse about individual rights and the obligation of government to protect them, we lack a language to talk about harm to the community. Steelworkers expressed this in the course of the litigation connected with the closing of U.S. Steel's Youngstown works. We tried to invoke the idea of a "community property right" that is violated when a company comes into a community, dirties its air, fouls its water, asks for and receives tax breaks and other benefits to smooth its way, sucks up the energies of the community's young people for generations, and then tries to throw the community away like an orange peel and walk off. But under existing law, what is damaged in this situation is not a right and not property. Harm to the community is a fact that does not give rise to enforceable claims.

The political philosophy of possessive individualism cannot help us, but perhaps the experience and tradition of the labor movement can. The most sacred concept of the labor movement is solidarity, or, more fully, that "an injury to one is an injury to all." This sense of connectedness, of choosing what benefits all of us rather than what helps one and hurts another, is the fundamental reason for brownfield reindustrialization.

The greenfield model of modernization reminds one of how the rebuilding of cities was envisioned 25 years ago. At that time it was supposed that the best way to rebuild a city was to bulldoze areas several square blocks in size, disperse the residents to the four winds and completely replace the housing stock. Only gradually did it become clear that this was a crude and in the long run self-defeating approach, tending to replace old slums with new ones. The newer vision of how to remake a city is to rebuild the structures gradually without relocating the people or destroying the social fabric of neighborhoods. Let the churches, settlements, and traditional meeting places of the community remain. Begin to rebuild in the least densely populated parts of an area. As new housing becomes available on the first micro-sites, relocate residents *within* the neighborhood. Proceeding in this fashion with, so to speak, a scalpel rather than a sledgehammer, cities can be rebuilt without losing thier social identity.

In 25 years, no doubt too late for many, many Youngstowns and, perhaps, Pittsburghs, this is how every civilized nation will also be modernizing its industry. Several of us in Youngstown had a glimpse of that future when we talked with the editor of the journal of the Swedish metalworkers' union, Per Ahlström. He began by emphasizing that Sweden, like the United States, is a capitalist economy. Then he went on to describe the Swedish steel crisis and how it was resolved. Several years ago, he said, Sweden faced the same problems of overcapacity and low profitability that now exist here. There were three Swedish steel companies, two privately owned and one owned by the government. Each was trying to carry on the whole steel-making process from blast furnace to rolling mill, and all were losing money. Accordingly, the Swedish government insisted that the three enterprises rationalize their activities. At the same time, however, it was decided as a matter of principle that rather than concentrate all steel-making in a single location, it would be socially preferable to preserve each of the three traditional steel towns if a way could be found to do so. The resolution was that each company remained where it was, but each was henceforth responsible for a single phase of steel-making. The mill closest to sources of iron ore in northern Sweden did the initial processing. The mill located on the seacoast did most of the finishing, and so on. Meantime, because all modernization and rationalization tends to eliminate jobs, imaginative programs were designed to help people leave the steel industry, not in shock and defeat, but with a sense of moving forward in their lives. Swedish employers were required to list all job openings, and a computerized printout was posted each day in the mill itself. Persons desirous of visiting other communities where there were job openings were paid to do so, as were their spouses. Every steelworker was guaranteed two years' pay during the period of transition. The social objective, our visitor stated, was that no one ever be compelled unwillingly to leave a job.

Sweden does what it does for essentially "political" reasons. In an article in *The Nation,* Helen Ginsburg (1980) quotes an unnamed Swedish official: "Swedes are not particularly religious but one thing we do hold almost sacred is everybody's right to work." The result, she continues, is that the unemployment rate in Sweden was 1.7 percent from 1960 to 1970, and 2.1 percent from 1971 to 1979. This is not because the Swedish economy in general or its steel industry in particular are immune to the shocks affecting other capitalist economies. On the contrary, Sweden is more dependent than the United States on exports, and has no coal or oil of its own.

> The answer is that the Swedish commitment to full employment is politically unassailable. Even though it has traditionally been regarded as an important means of raising the output of goods and services, and hence living standards, it is not viewed solely in economic terms. It is also linked to other vital goals. . . . [T]he concept of "normalization" is fundamental to the Swedish social welfare system; that is, the goal is to enable everyone to live as normal a life as possible and "to reduce the risk of isolation, loneliness and alienation." And work is considered the key to normal life. In short, a job is considered a basic right [Ginsburg, 1980].

Listening to Per Ahlström in Youngstown, Ohio, was like hearing a fairy tale. For instance, early retirement, which is the *objective* of the United Steelworkers of America in its collective bargaining about shutdowns, is, according to Ahlström, considered a *defeat* in Sweden because it deprives a person of years of contribution to society as a worker.

The principle is not that industry should always be modernized where it currently exists. A company might carry out brownfield industrialization in such a way as to sacrifice social values to profit, as General Motors has in Detroit, where the new Cadillac assembly plant they are building will destroy 1021 homes and apartment buildings, 155 other businesses, several churches, and a hospital, and displace 3500 people, thereby eradicating a traditional and racially integrated neighborhood.[8] The principle is that economics and technology should be subordinated to the preservation and nurturing of community. This principle may be expressed in the presumption that it will ordinarily be preferable to rebuild in one place, rather than to scrap and move on.

NOTES

1. In 1977, the Federal Trade Commission found that for the period of 1961-1971 the United States had the highest profit rate, Japan the second highest, and the European Community the lowest, when profit was measured by net income as a percentage of sales.

When profit was measured by net income as a percentage of equity, the profit rates of the United States and Japan were approximately equal, and that of the European community was again the lowest (Federal Trade Commission, 1977: 504-505). In 1980, the U.S. Office of Technology Assessment (OTA) reported that in the period of 1969-1977 net income as a percentage of net fixed assets in five major steel-producing countries was as follows: United States, 6.7 percent; Japan, 1.7 percent; West Germany, 2.9 percent; United Kingdom, -5.3 percent; France, -8.3 percent (U.S. Office of Technology Assessment, 1980: 126).

2. The Jones Conable bill, H.R. 4646/S.B. 1435, would permit industry to accelerate depreciation of new investment for tax purposes. At present, investment is depreciated over an estimated useful life of approximately fifteen years. Jones-Conable would permit depreciation of buildings in ten years, with 70 percent taken in the first five years, and depreciation of machinery in five years, with 52 percent taken in the first two years. Congressman Vanik (D-OH) has estimated that this bill would cost the federal government $122 billion in taxes over the first five years (Congressional Record, November 27, 1979).

3. A "tonne," or metric ton, is 2204.6 pounds.

4. Cost data for the Conneaut plant are confused for several reasons: First, the company projects two (privately, four) stages of construction; second, capital costs have increased dramatically since U.S. Steel first made public statements about the proposed plant; and third, estimates are sometimes made per ton of raw (liquid) steel and sometimes per ton of finished (shipped) steel. The figure given was provided by Edgar B. Speer, then chairman of the board, in 1976 (Industry Week, April 15, 1976).

5. The question may arise then, why should a U.S. Steel prefer greenfield expansion? The answer appears to be that once a greenfield plant is built, it can produce steel more cheaply than a modernized brownfield facility. Thus if the company can induce the government (that is, the taxpayer) to build the plant for it by means of tax incentives, this becomes the desirable option for the firm.

6. In Bucks County, Pennsylvania, where U.S. Steel's Fairless Works is the largest employer, Andrew Moody of Chase Econometric Association has forecast that a gradual shutdown of the works over a two-year period would result in the loss of 8000 jobs at Fairless, 1600 jobs in other manufacturing industries, and 11,000 jobs in non-manufacturing areas, with a resulting unemployment rate in Lower Bucks County of 10-15 percent, just as in Youngstown (Wolf, 1980).

7. Williams is the principal author of a federal Department of Energy report on methods to use for computing the socioeconomic impact of large industrial projects.

8. Karl Greimel, dean of the Lawrence Institute of Technology School of Architecture and an experienced industrial architect, testified in court that the plant could be built in a much smaller space so as to save most of the "Poletown" neighborhood. For instance, instead of placing a mammoth parking lot adjacent to the plant, General Motors could build a multilevel parking structure or provide parking on the plant roof. General Motors has refused to make such changes (Moberg, 1981; see also Serrin, 1981).

REFERENCES

Argus Research Corporation (1977) "Steel: an industry in flux." Report, August 31.
BRADFORD, C. A. (1977) "Japanese steel industry: a comparison with its United States counterpart." (mimeo, June 24).
Business Week (1979) "Big steel's liquidation." September 17.

Federal Trade Commission (1977) The United States Steel Industry and Its International Competitors: Trends and Factors Determining International Competitiveness. Washington, DC: Government Printing Office.

GINSBURG, H. (1980) "A national commitment: full employment the Swedish way." The Nation (December 6).

HIRSCHHORN, J. (1980) "Putting steel into steel." New York Times (December 17).

Iron and Steel Trades Confederation (1980) New Deal for Steel. London: Penshurst Press.

Local 1330 USWA v. U.S. Steel Corp. (1980) 492 F. Supp. 1 (N.D. Ohio).

LYNCH, R. (1981) "Reagan campaign themes are now going south." In These Times (March 18-31).

MOBERG, D. (1981) "Detroit: I do mind moving." In These Times (February 4-10).

New Haven Advocate (1979) "U.S. Steel closing down." December 5.

New York Times (1981) "The turnaround at U.S. Steel." February 19.

Policy and Management Associates, Inc. (1978) Socioeconomic Costs and Benefits of the Community-Worker Ownership Plan to the Youngstown-Warren SMCA. Cambridge, MA: Author.

RODERICK, D. (1979) "Is there an OSEC in our future?" Speech presented to the Purchasing Management Association, Houston, September 11.

ROHATYN, F. (1981) "Reconstructing America." New York Review of Books (March 5).

ROWTHORN, B. and T. WARD (1979) "How to run a company and run down an economy: the effects of closing down steel-making in Corby." Cambridge Journal of Economics (December).

SERRIN, W. (1981) "Huge new G.M. plant, like many, to get subsidies." New York Times (February 25).

United States v. Bethlehem Steel Corporation (1958) 168 F. Supp. 576, 616 (S.D.N.Y.).

U.S. Council on Wage and Price Stability (1977) Report on Economic Conditions Within the American Steel Industry. Washington, DC: Government Printing Office.

U.S. Office of Technology Assessment (1980) Technology and Steel Industry Competitiveness. Washington, DC: Government Printing Office.

U.S. Steel (1979a) Annual Report. Pittsburgh: Author.

———(1979b) Highlights of U.S. Steel Corporation's Proposed Lakefront Plant. Pittsburgh: Author.

WOLF, D. (1980) "Fairless Works: what's on road ahead?" Bucks County Courier Times (December 2).

4

PROBLEMS OF ECONOMIC DETERIORATION

Bennett Harrison
Barry Bluestone

From one end of the country to another, in the older industrial areas within the so-called Frostbelt from Maine to Michigan, and—surprising as it may seem—in the southern and southwestern Sunbelt as well, workers and their families are experiencing a wave of job losses due to the shutting down of factories, stores, and offices. Some of these shutdowns involve the often unavoidable failure of small, locally owned firms that are unable to survive in the unending competitive struggle with other businesses.[1] Sometimes, large corporations close one of several of their branch plants or stores as part of a planned "rationalization" of the business. Yet another category of closings is associated with the managerial strategies of giant conglomerates, which often buy up previously independent businesses, milk them for the cash they can generate, and then dump them or close them down when other, more profitable ventures become available. And finally, multiregional or multinational corporations sometimes close down a business in one place expressly to relocate it.

The objective of this chapter is to combine what is known about these various kinds of plant closings and relate them to the larger and more fundamental issues of how capital in general is "moved," what social costs are incurred in this movement, and what might be done to ensure more equitable and balanced economic growth.

ECONOMIC DEVELOPMENT, CAPITAL MOBILITY, AND THE CURRENT CRISIS

To make sense of this problem, some basic features of the ways in which a capitalist economic system is organized should be reviewed. In *any* kind of economy, workers need capital goods to be socially productive. A worker's productivity is directly related to his or her access to capital goods.

In our (so-called) free-enterprise system, workers as a class neither own nor control capital to any significant degree. The people who do have every incentive to exercise their control with the objective of making as much profit as they can and accumulating as much wealth as possible. To meet this objective, employers must keep their costs of production down and coax as much productivity out of their employees as available technological conditions allow.

Employers and managers use a variety of methods to try to hold down labor costs. One way is to resist, or at least restrain, wage and benefit increases. Periodically, they also try to reorganize work rules and procedures to make them more efficient, especially in conjunction with the introduction of labor-saving machinery. This machinery is used to produce more output with the same work force or the same output with a smaller or less skilled work force.

Historically, workers have often responded to management's attempts to restructure wages, working conditions, and the location of work unilaterally by organizing to protect themselves and their families. Through labor unions, they have sometimes been able to win (or at least protect) some degree of job security, a measure of control over day-to-day work rules, and higher wages and benefits such as pension rights.

The list of what labor counts as victories (often called the "social wage") over the last sixty years is well known: unemployment insurance benefits, pension rights, public assistance, insurance against loss of income due to job-related disabilities, legal minimum wages, occupational health and safety regulations, and equal employment opportunity laws.

The true dollar cost of labor to the firm is the sum of ordinary wages (or salaries), company paid benefits, and employer contributions to the social wage. An increase in any of these labor costs often comes at the expense of profit.[2]

Most employers see these gains for labor, especially those that have resulted from government regulations, as unwarranted interference with their right to run their own businesses as they please. How can they successfully oppose popularly supported market intervention? Given that they, and not the workers, own the buildings, machines, and raw materials needed for production, the mere threat of withholding capital services can impose such havoc on a community that it is forced to make substantial concessions to them. The laws of private property make the ultimate threat possible—to deny or *move* those capital goods and services altogether.

A second reason for this shift of capital arises from the nature of competition in a global economy. The recent trend has been toward an increased concentration of economic and political power in the hands of a steadily declining number of giant corporations. These huge firms have the power and means to expand and manage their operations over long distances, both interregionally and internationally.[3]

Ironically, instead of helping to augment their control over the market, the expanding international arena has created an intense competition among these giants. This *forces* them to go as far afield with their capital as their abilities and international politics will permit, especially in search of more highly exploitable labor.

THE CONTRADICTORY NATURE
OF CAPITAL MOBILITY

Economically speaking, efficient allocation of resources in a full employment economy would (in theory) create the greatest chance for labor to realize its productive potential.

Acceptance of this principle, however, in no way implies that minimizing regulatory controls on private investment is the best way to avoid the inefficient use of resources. In fact (as will be shown later), just the opposite may be true.

There may be a need for public leverage over private capital. It is easier to understand why this may be true if we examine what businesspeople mean by the term "efficiency" as contrasted with what economists mean by "efficient allocation."

In a so-called free-enterprise economy, decisions over capital investment are made on the basis of private profit. To ensure maximum profits, entrepreneurs must use their investment funds in the ways that bring the highest rate of return. This essentially is business's definition of efficiency. From an economist's point of view, the productivity of labor, land, and "social infrastructure" are no less important than the productivity of physical investment.[4] Wasting labor or land, or old but still useful buildings, is just as economically inefficient as wasting financial capital. The efficient allocation, or use, of *all* resources is summarized in the notion of "social efficiency." According to Bohm (1973: xiv), "Social efficiency" involves "an attempt to take into account *all* individuals' evaluations of *all* consequences of economic acts." It involves the "efficient use of [all of] a nation's resources."

Thus capital mobility, which leaves productive labor unemployed, is not necessarily *efficient* even if the reassignment of capital permits the firm to post higher earnings. The loss in the productivity of labor can mean that social efficiency—society—has suffered. What is efficient from a private business perspective can thus involve extraordinary prodigality from the perspective of society.

Society suffers when capital mobility results in "social inefficiency" because society is absorbing costs that should be absorbed by business. These costs are not included in what business must cover to make profits; thus posted profits are higher than they should otherwise be. These societal costs are called "negative externalities"; evaluations of capital reallocations should consider the possible creation of these "negative externalities."

THE EXTENT OF THE PROBLEM

The states currently attracting investment relocations are in the West and the Sunbelt. The states suffering most from disinvestment are those in which manufacturing plants have long made up the heartland of organized labor: Wisconsin, Michigan, Illinois, Indiana, Ohio, New York, New Jersey, and Pennsylvania.[5] The correlation between these geographical patterns of capital investment and the employment shifts shown in Figure 4.1 is unmistakable.

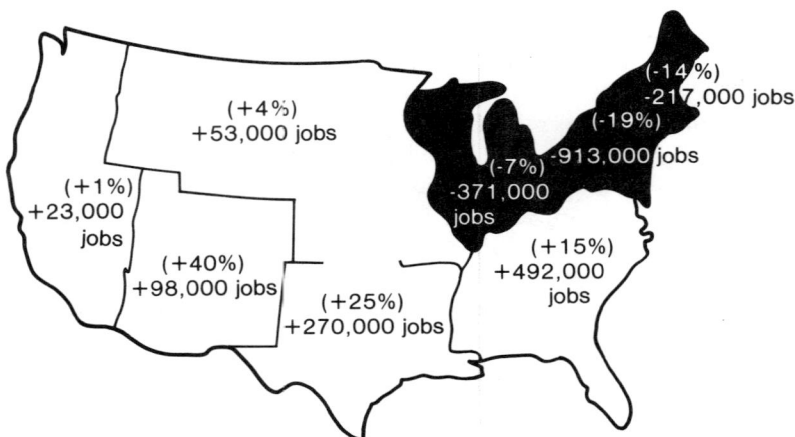

SOURCE: **Dollars and Sense Magazine,** April 1979.

Figure 4.1   Manufacturing Jobs Lost and Gained, 1967-1976

Capital is also moving out of the country. Since the end of World War II, the growth of American investment in other countries has been enormous. Between 1950 and 1974, this foreign investment increased *ten times*, from about $11.8 billion to over $118 billion.[6] In the same period, gross private domestic investment grew at *less than half* that rate, from $53.8 billion in 1950 to $215 billion in 1974 (Council of Economic Advisers, 1977).

The problem, then, is the *disinvestment* that results from capital reallocation: the running-down or liquidation of the net productive capacity of a business in one location for the purpose of moving the capital to another location. Private disinvestment in any particular area inevitably displaces people, machinery, and land. If the displacement is great enough, sudden enough, and sufficiently concentrated in one place, the entire community may feel the blow. Other firms and their employees will also be affected. The tax base can erode as well, with all the threatening implications that holds for public-sector workers in the area.

Yet, even with enormous capital flows, capital reallocation or "mobility" per se is not the enemy. In a world of growing material scarcity, one realizes that resources must be allocated wisely and thoughtfully. This often calls for removing capital from some types of activity to use it more productively in others. The real problems are the accelerating velocity of capital—its increasingly frenetic, unplanned, and abrupt movement—and the fact that the real social and economic costs (negative externalities) of its relocation hardly ever enter into the decisions of corporate managers. The basic issue, then, is to assure that the decision whether or not to transfer capital from one use or location to another will include the costs of worker and community dislocation.

HOW CAPITAL "MOVES OUT" OF OLDER AREAS[7]

Capital can be moved literally and physically by closing down operations at one location, selling the old plant, and transporting the equipment to a new site. There, a new building is constructed and operations are resumed. One such shift occurred just after World War II, when United Aircraft's Chance-Vought division moved from Bridgeport, Connecticut, to Dallas, Texas. This relocation was one of the most spectacular migrations in recent industrial history. In all, approximately 1,500 people, 2,000 machines, and 50 million pounds of equipment were involved.

This type of capital movement, although dramatic, is not very common. Using data from the Dun and Bradstreet Co., a private business credit rating service, David Birch of MIT concluded that, between 1969 and 1976, only about 2 percent of all private-sector annual employment change in the United States was the result of *physical relocation or migration* as in the case of Chance-Vought. During the 1969-1970 recession, the rate was even lower—perhaps 1.5 percent per year. This last is a famous statistic that has been quoted time and again by academic researchers[8] and by political conservatives seeking to prove that the growing attempts to restrict capital flight and to make business more accountable to the public "are solutions to a non-existent problem" (McKenzie, 1979: 5).[9]

But the problem of capital mobility or "capital flight" cannot be so easily trivialized for two reasons. First, Dun and Bradstreet undoubtedly overlooks many short-term *relocations* by classifying them as *closings* in the old location and *openings* in the new, instead of *migrations.* For this reason the Dun and Bradstreet data provided only a *minimum* estimate of capital flight.

THE MANY OTHER FORMS
OF CAPITAL MOVEMENT

The second reason for the underestimation of capital flight arises because there are many "subtle" ways of disinvesting, equivalent to migration as such. Some of these ways are listed below:

(1) Both large and small companies may run down their older facilities (for example, by not replacing worn-out machinery), and use the savings in the form of depreciation allowances to reinvest in other branches of their own firms, in other people's businesses, or perhaps in municipal bonds.

(2) Companies may take that last step and close the older facility altogether. Land and/or buildings might be put on the market, and the machinery sold for scrap to other branches or to other firms. If the owner formally declares bankruptcy, Chapter XI of the bankruptcy laws provides a procedure that enables even small businesspeople to retain a substantial surplus from their creditors, which may then be reinvested in new economic activity or used as a kind of pension. The latter practice is common among small entrepreneurs who want to retire but have no heirs with whom to leave their business or who can find no one who wants to run it.

(3) Multiplant, multistore, and multioffice corporations may gradually shift some machinery, skilled labor, managers, or simply marketing responsibilities from their older to their newer facilities located in some other city, state, or country. The old facility remains in operation, at least for the time being, but at a lower level of activity.

(4) The multibranch corporation may not, in the short run, physically remove any of the older plant's capital stock, but will instead reallocate profits earned from the plant's operations to its newer facilities for investment. Such "milking" of a profitable plant is especially common among conglomerates, whose managers have sometimes described their acquisitions as "cash cows." This is the most subtle and invisible form of disinvestment.

In all four cases some form of capital is reallocated from one plant, office, store, warehouse, or hospital to another. Sometimes the transfer takes place among the branches of a single firm and sometimes it occurs

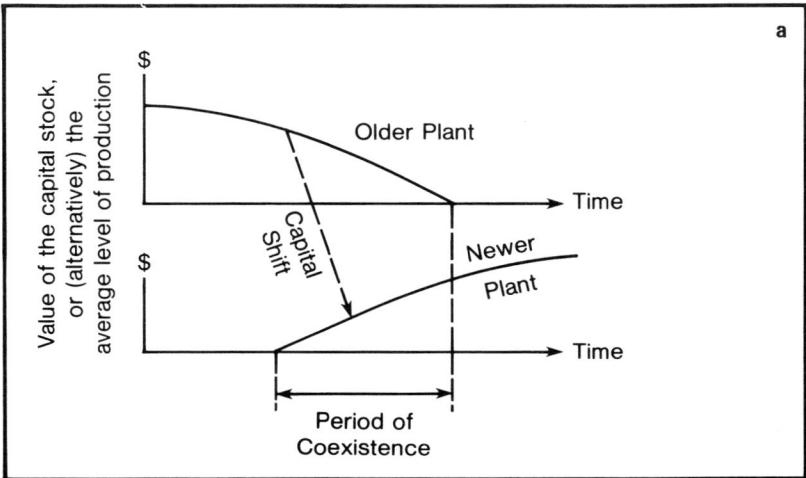

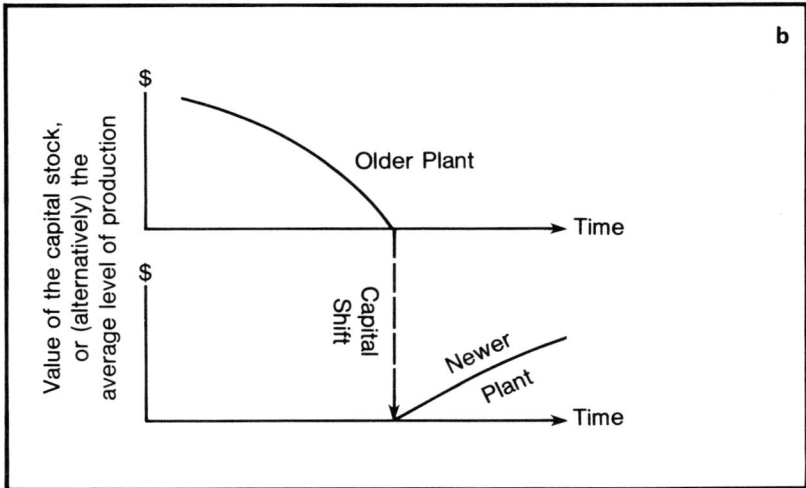

Figure 4.2 Possible Types of Capital Shift Between Business Establishments over Time

across firms, possibly mediated by a bank, insurance company, or local public development authority. In other cases, the transfer takes the form of finance capital (profits or savings reinvested elsewhere). In only a few cases does it consist of actually transferring capital goods: machinery, equipment, tools, or parts of buildings.

Regardless of the particular form of capital involved, in nearly all cases, the "old" and "new" facilities *coexist* for at least a short period of time. Often, the older facilities are eventually closed down completely, particularly after the newer ones become fully operational. But the timing can vary enormously. Some giant corporations such as General Electric have operated parallel plants for thirty years, while conglomerates such as Genesco, Sheller-Globe, and Lykes may "milk" some of their acquired subsidiaries quite rapidly. In numerous instances, plants have been run down and eventually closed within three to ten years of the date of acquisition.

This process of capital shift, depicted in part a of Figure 4.2, can occur directly or indirectly, within or between firms, but always from older, declining facilities to newer, growing ones. The two plants, stores, or offices can actually coexist for a period of time. In fact, the old plant may *never* be shut down entirely.

In special cases, depicted in part b of Figure 4.2, the "old" shutdown and the "new" opening occur more or less simultaneously, or at least in rapid sequence. This kind of "instant migration" of a plant is obviously only a variation of the more general type of capital shift. And, not surprisingly, because money, time, and effort are necessary to shift capital or at least capital goods across space, relatively few shifts are of this rather extreme kind.[10]

However, because operating a multibranch firm is invariably more costly and difficult than running a single-branch enterprise, small businesses with less capital resources are most likely to make simultaneous or rapidly sequential moves when they do actually relocate. Researchers have found that the majority of these "relocating" small businesses have a single plant and fewer than twenty employees. This type of relocation has caused serious problems for workers in the textile, apparel, shoe, and electronics industries.

Our conclusion is that the extent of capital migration and disinvestment is substantially underestimated. The first step toward solving the resultant societal problems is an acceptance of a more realistic level of its occurrence.

CAPITAL MOBILITY, LABOR MOBILITY,
AND SOCIAL JUSTICE

To repeat: Capital mobility per se is not at fault; the problem is its uncontrolled velocity. Finance capital can be transferred by Telex and products can be transported by jet freighters. People move more slowly as most have roots in their particular communities. Moreover, in the words of Nobel Prize-winning economist Paul Samuelson (1958: 337): "People want to improve their communities, not abdicate from them." Yet, unrestricted, rapid capital flow destroys communities and personal assets while it creates an industrial "refugee" crisis of substantial magnitude—whole subcommunities of workers without jobs.

Conceivably, in the very long run, most resources can find new productive uses of one kind or another—including labor. But the often unasked question is, *How long?* Displaced workers and abandoned buildings might be reemployed eventually in a dynamic economy; but how long is "eventually"?

Social costs (externalities) must be taken into account in determining the appropriate speed of capital mobility in this day and age. The problem is to design regulations so this appropriate velocity is maintained and results in a technology neither too small nor too large, but appropriate for the job to be done (Schumacher, 1974). The appropriate velocity for capital reallocation would be difficult to pinpoint, but it would surely be less than what the unregulated, private market would dictate.

Another problem concerns the distribution of the burden—the issue of social justice. Even if the outcome of capital mobility shows a net gain in overall productivity, some people stand to lose considerably. The U.S. textile worker loses while the Taiwanese worker gains; the nonunion southern auto worker takes a job away from a union worker somewhere else. And in every case of capital shift, owners gain—at least in the short run—at the expense of *some* workers somewhere.

THE STANDARD POLICY RESPONSE

The almost sacred nature of private capital in our society largely precludes most direct responses to the economic waste and human suffering caused by capital flight. At best, liberal government policy is aimed at ameliorating the effects rather than attacking the causes of dislocation.[11]

The worker who is displaced because of some investment decision beyond his or her control is often eligible for a whole range of temporary income maintenance programs, from unemployment insurance benefits, trade readjustment assistance, and pension reinsurance to AFDC, emergency assistance, and food stamps.

Unquestionably, these programs are necessary and valid, but the weakness of the welfare solution is that it attempts to deal with what is basically an allocation issue with simple government redistribution; practically none of the costs are rerouted to the business that stands to gain most from the initial capital shift.[12] Under present rules there is no disincentive to move one's plant even if the social costs outweigh private gain by a wide margin.

Thus welfare is an insufficient solution to the capital problem. It fails to reuse abandoned buildings and land, it does not give people what they really want—jobs—and it forces taxpayers to pay for costs created by private corporations, which have wealth and power that far exceed that of the people being displaced. A better solution would take direct aim at the investment decision itself.

Various alternatives may be feasible: mandated employer-paid severance pay, relocation and retraining allowances, a *reversal* of government tax regulations and laws that actually promote capital flight, and worker or community buyouts of the enterprise.

The operative myths must be stripped away and the disturbing reality concerning capital mobility must be clearly confronted. Private profit should not continue to be the *sole* criterion for corporate investment decisions that can have a significant negative impact on workers, communities, and the American economy as a whole. A public balance sheet, which weighs the considerable costs to people, towns, and public treasuries of decisions made by conglomerates and other corporations in their own interests, should replace the traditional private balance sheet as the standard for decisions concerning capital mobility.

NOTES

1. It must be noted, however, that many small business owners *choose* bankruptcy as a way to recoup some of their invested capital when they or their offspring grow tired of the business.

2. In many instances, particularly cases in which industry enjoyed monopoly advantage, the costs of the social wage can be passed through to consumers in the form of price increases or shifted back onto workers in the guise of lower money wages. In either case, profits may be maintained. The issue of who pays for the social wage is decided in the long run by the traditional struggle between capital and labor.

3. The best introductory studies are still Hymer (1970) and Barnet and Müller (1973). This does not mean that small firms do not also engage in overseas production—they do (especially in apparel and electronics). But by far the most important agent of international capitalist development is the giant multinational corporation.

4. "Social infrastructure" is the term economists use to include such factors of production as roads, rail transport, water, sewage, and power, whose availability reduces production costs for industry. Social infrastructure may be either publicly or (under certain circumstances) privately supplied.

5. Data from the First National Bank of Boston's newsletter for Fall, 1978, reveal that—were it not for the computer industry—New England would have been in the same category during these years of transition.

6. Kelly (1977) cites these official U.S. government statistics, and observes in passing that the total overseas output of American multinational corporations is now greater than the gross national products of every country in the world except the United States and the Soviet Union.

7. In this chapter, we are interested mainly in private disinvestment in older regions and its consequences. But of course, capital also moves into such areas continually. It is simply incorrect to imagine the situation as one in which *everything* is leaving. New York City, Newark, and Cleveland are hardly becoming ghost towns! One problem is that the businesses that still operate there, or the new ones that are opening, are growing too slowly to compensate for those businesses that are cutting back, closing, or leaving. Moreover, the new companies may not want to employ the people who need the new jobs most: those who have been thrown out of work by the old closings. The jobs that make up the new economic base may pay lower wages and offer less steady or fewer hours of work than did the old base. We estimate, for example, that it takes 1.5 to 2 service sector jobs to make up in family income what is lost by the elimination of one manufacturing job. For details, see Bluestone and Harrison (1982: chap. 4) and Harrison (1984).

8. See Jusenius and Ledebur (1977a) and Weinstein and Firestine (1978). In earlier published work, one of us also cited the 1.5 percent figure as evidence that there is relatively little short-term spatial movement of businesses in the United States (Harrison and Kanter 1978).

9. The basic data may be found in Birch (1979). Birch's work has been quoted by the *Wall Street Journal* and by many southern researchers to counter the argument that southern growth is occurring at the expense of northern workers. Research papers by regional economists that have cited Birch's famous "one and a half percent" have been similarly used. For example, one study of plant migrations out of the state of Ohio over the period 1970-1974 found that "migrant firms were few in number and mostly small in size" (Jusenius and Ledebur 1977b). When this conclusion was announced in late 1977, it was quickly reported by daily newspapers all across the state, just when a coalition of trade unionists, community activists, and church leaders was trying to find a way to save the jobs of the 4,100 workers who had been laid off that September when the Campbell works of the Youngstown Sheet and Tube Steel mill was closed down by its conglomerate parent, the Lykes Corporation. Another study, by two Harvard Business School economists, was published earlier this year in the prestigious journal of the Wharton School of Finance (Leone and Meyer, 1979). Leone and Meyer (1979: 23) report that "of the more than 1450

Northeast manufacturing facilities that can be definitely identified as making a long distance move from 1971 to 1976 . . . only 128 emigrated to the Sunbelt."

10. This has not stopped business propagandists from continually *threatening* to move out of town if they do not get their way. In an interview with the trade magazine *New England Business* (October 1, 1979: 14), Associated Industries of Massachusetts lobbyist Jim Sledd asserted that Massachusetts high-technology firms are "very mobile" and that "they can be on the back of a flatbed truck tomorrow and be in North Carolina if things don't go the way they want them to."

11. Worse yet, in providing welfare, the government becomes handmaiden to the disinvestment process by socializing the costs generated by the firm. For the most part, this means that workers who retain their jobs are taxed to subsidize those who have been forced out of theirs.

12. It is perhaps important to recognize that under current market arrangements, shifting the burden of social costs nominally onto capital (through severance pay taxation, for example) does not guarantee that capital ends up paying the cost. Unfortunately, firms operating with monopoly power are often in the position to shift taxes forward onto consumers in the form of higher prices. For capital to bear the full social costs of privately profitable plant closings, it would ultimately be necessary to at least partially regulate prices as well. Hence the need to couple industrial policy, regulation of capital mobility, and "incomes policy" in any sort of reindustrialization program that has a real chance of succeeding.

REFERENCES

BARNET, R. and R. MÜLLER (1973) Global Reach. New York: Simon & Schuster.
BIRCH, D. (1979) The Job Generation Process. Cambridge: MIT Program on Neighborhood and Regional Change.
BLUESTONE, B. and B. HARRISON (1982) The Deindustrialization of America. New York: Basic Books.
BOHM, P. (1973) Social Efficiency: A Concise Introduction to Welfare Economics. New York: John Wiley.
Council of Economic Advisers (1977) 1977 Economic Report of the President. Washington, DC: Government Printing Office.
HARRISON, B. (1984) "The economic transformation of New England since World War II," in L. Sawers and W. K. Tabb (eds.) Sunbelt-Snowbelt. New York: Oxford University Press.
———and S. KANTER (1978) "The political economy of state job-creation business incentives." Journal of the American Institute of Planners (November).
HYMER, S. (1970) "The multinational corporation and the law of uneven development," in J. M. Bhagwati (ed.) Economics and World Order. New York: Macmillan.
JUSENIUS, C. and L. LEDEBUR (1977a) "A myth in the making: the southern economic challenge and northern economic decline," in E. B. Liner and L. K. Lynch (eds.) The Economics of Southern Growth. Durham, NC: Southern Growth Policies Board.
———(1977b) The Migration of Firms and Workers in Ohio, 1970-75. Columbus, OH: Academy for Contemporary Problems.
KELLY, E. (1977) Industrial Exodus. Washington, DC: Conference on Alternative State and Local Public Policies.

LEONE, R. A. and J. R. MEYER (1979) "Can the Northeast rise again?" Wharton Quarterly (Winter).

McKENZIE, R. B. (1979) Restrictions on Business Mobility. Washington, DC: American Enterprise Institute.

SAMUELSON, P. (1958) "Aspects of public expenditure theory." Review of Economics and Statistics (November).

SCHUMACHER, E. F. (1974) Small Is Beautiful. New York: Perennial.

WEINSTEIN, B. L. and R. E. FIRESTINE (1978) Regional Growth and Decline in the United States: The Rise of the Sunbelt and the Decline of the Northeast. New York: Praeger.

5

ABSENTEE OWNERSHIP, INDUSTRIAL DECLINE, AND ORGANIZATIONAL RENEWAL

Christopher Meek
Warner Woodworth

Today, old industrial cities of the northeastern and midwestern United States face the distressing possibility of becoming the modern industrial version of the ghost towns that followed the mining booms of the 1800s. If this does happen, industrial deterioration will be the principal reason, for industry in these areas has been on the decline ever since the postwar boom of 1947 (O'Leary, 1974; Srb, 1975; Foltman et al., 1975; Chinitz, 1977; Foltman and McClelland, 1977a). Since 1966, New England, the mid-Atlantic, and the Great Lakes states have together lost close to 1.4 million jobs due to plant relocations alone (Cook, 1980: 38). During the twenty years that elapsed between 1953 and 1973,

> manufacturing jobs declined 23.4% in New York, 10.5% in Pennsylvania, 8.6% in Connecticut, 2.3% in New Jersey but increased 71.6% in North Carolina. Nationally, manufacturing employment rose by 13.8% from 1953 to 1973 [O'Leary, 1974: 5-1].

As a case in point, the above-cited statistics illustrate that New York has been extremely hard hit by the problem of manufacturing decline. Between 1953 and 1974, industrial employment in that state alone dropped by 537,700 jobs, from a total of 2,118,900 to 1,581,200 (Foltman and McClelland, 1977b: 1). These figures pale, however, when compared with what increased employment might have been had New York maintained a growth rate on a par with the national average. When job losses due to plant closings and manufacturing cutbacks are combined with the declining industrial growth rate in New York State for only the ten years between 1965 and 1975, lost employment soars to a figure in excess of 2 million (Foltman and McClelland, 1977b). As a result of these problems, New York's position of strength as one of the nation's major industrial states has been seriously weakened. By 1975 it had dropped from a peak of employing 12.8 percent of the U.S. industrial work force to a level of 7.7 percent (Foltman et al., 1975; Chinitz, 1977: 57).

Clearly, these trends reveal a disturbing predicament. Not only are industrial cities in states like New York plagued by heavy employment losses as a result of plant closings and manufacturing cutbacks, but they are also failing to recover these losses through either new or expanding industry. They cannot keep the operations they have, nor can they convince new companies to locate within their boundaries. Thus responding to this predicament, if it is still possible, will require that communities look to their own existing industrial base and find ways to "revitalize" what they already have. It is this question of revitalization that is the primary concern of this chapter. It will be helpful first to consider the impact that the gradually eroding base of local influence over industry has had upon the encroaching problem of industrial decline.

ABSENTEE OWNERSHIP AND COMMUNITY DECLINE

Now this red line of cleavage, in material interest and sentiment, runs not between those who own something and those who own nothing, as has habitually been set out in the formulas of the doctrinaires, but between those who own more than they can personally use and those who have urgent use for more than they own. The issue now is turning not on a question of ownership as such, but on absentee ownership [Veblen, 1923].

Typically, the problem of manufacturing deterioration (for example, plant shutdowns, the outmigration of industry, and declining industrial growth) is blamed on high taxes, high labor costs, and excessive work stoppages. Research has illustrated that there is validity to the first two assertions, but that the third is unfounded (Lindsay, 1975; Beckman, 1974). It has been shown that lost time due to work stoppages in New York, for instance, is not particularly greater than in other industrialized regions of the United States. More important, blaming the problem of industrial decline on unionized workers and the government serves only to divert attention from some of the significant contributions that industry itself has made to the problem of industrial deterioration. More specifically, it is important also to examine the role that increased absentee ownership and control over local economic resources has had in relation to the problems of northeastern manufacturing communities.

The impact of absentee ownership and control of the local economy has for many years been a topic considered in studies concerned with the structure of power and decision making in communities. During the early years of community studies, most theoretical and empirical investigation in this area centered on the question of whether communities are controlled by small groups of elites (Lynd and Lynd, 1937; Hunter, 1953; Vidich and Bensman, 1958; Clelland and Form, 1964) or are actually more pluralistic in structure (Dahl, 1961; Sayre and Kaufman, 1960; Wildavsky, 1964; Smith, 1965). In the process of seeking the answers to these questions, it became increasingly apparent that a variety of factors influence the structure of community power and decision making. Thus the study of community power and decision making gradually gained sophistication and adopted a situational approach (Clark, 1967). Among the many factors that were identified as having sigificance in determining who makes which community decisions and what criteria determine these decisions was the degree of absentee ownership and control over the local economy.

Probably the first study to consider critically the influence of absentee ownership on community structure and relationships was Warner and Low's fourth volume in the Yankee City series, *The Social System of the Modern Factory* (1947). While conducting their classic field study of the community life in Newburyport, Connecticut, Warner and Low became deeply impressed by a citywide strike that closed seven operations in the community's principal industry, shoe manufacturing. This collective action not only obtained worker backing but also garnered support from much of the local community, including the mayor and city merchants. Although subsequent historical analysis (Thernstrom, 1965) has shown that Warner and Low drew many erroneous conclusions about the significance of the strike and the evolution of industry in Newburyport because of their overreliance on oral history and lack of analysis of historical records, they did hit upon some important insights into the question of absentee control and its impact on community that have since been elaborated further through subsequent research.

What Warner and Low found was that as outside ownership gained control over industry, this very important part of the Newburyport economy became increasingly less connected to the community. That is, as outside or absentee interests acquired control over the local shoe manufacturing industry, the management of these operations became further and further removed from the concerns and daily workings of the community. Gradually, the policies and actions of these organizations became disconnected from the needs of the local area and were geared primarily to the profit picture of the parent companies located elsewhere. Warner and Low (1947: 108-109) describe this change:

> In the early days of the shoe industry, the owners and managerial staffs of the factories, as well as the operatives, were residents of Yankee City; there was no extension of the factory structures outside of the local community. The factories were then entirely under the control of the community—not only the formal control of city ordinances and laws but also the more pervasive informal controls of community traditions and attitudes. There were feelings of neighborliness and friendship between manager and worker and of mutual responsibilities to each other and the community that went beyond the formal employer/employee agreement.
>
> With the vertical extension of the managerial hierarchy, the social distance between the top executives on the one hand, and the workers and community on the other, has increased to the point where these bonds of mutual friendship have virtually disappeared. Absentee ownership, which usually accompanies absentee control, accentuates this condition. The conflicts of interest between the two groups of owner/manager and worker/community thus become more pronounced because they lack the bonds of mutuality. Factory policies are established in the distant offices of large concerns; neither worker nor community now has any voice in them.

Thus the strike became a rebellion of sorts on the part of both the workers and the local residents against the powers that had come to determine the viability of the local economy and over which they no longer had influence. In cases in which the original manufacturers as citizens of the community had to respond and maintain some degree of reciprocity in their relationship with their employees and the rest of the community, the new managers and their companies were impervious to such influence and often so uninvolved in community affairs that the old, informal avenues the community had once used to influence industry had all but disappeared. Warner and Low (1947: 118-119) further illustrate this situation by comparing the relations of the managers of locally owned companies with those of the managers of the absentee-controlled operations:

> The vertical extension of its managerial hierarchy and the absentee control of factory C also made its relations with the community of Yankee City [Newburyport] much different from those of factories A and B. Typically, the managers and supervisory staffs of the locally owned, independent factories (like A and B) are native Yankees; frequently, they were born and grew up in Yankee City. They are thus involved in the general social life of the community, belonging to various associations, clubs, and other organizations. They subscribe to many general community attitudes which impinge upon their working relations, frequently causing them to modify working behavior away from that which would follow the single-minded dictates of the profit-making logic. Their life in the factory is not divorced from life in the community outside of working hours. Part of the motivation that determines their business behavior is the desire that fellow townsmen regard them as upright and fair businessmen who treat their employees properly. Such desires frequently militate against their acting logic; business advantages are sometimes sacrificed because the manager (or owner) places greater value on community prestige than he does on increasing factory earnings by some means which would endanger that prestige.

> In the case of absentee-controlled shoe factory C, however, there is little business need for the manager to participate in community activities. He can operate the factory in strict accordance with the orders of the main office more easily, in fact, if he does not take part, either as an individual or as a representative of the factory, in local associational activities. If he does involve the factory in community activities, he involves it and the larger enterprise of which it is a part in community responsibilities and subjects it to community pressure and control. This is precisely what the top officials of the large enterprise do not want. They want the factory to be as free as possible of community pressures so that its operations can be dictated in strict accordance with the profit-making logic. *This would even allow them, if it seemed desirable, to move their factory to another community.*

Since the Yankee City studies, a number of other research investigations have helped to clarify further how absentee ownership tends to alter the structure and nature of community relationships (Pellegrin and Coates, 1956; Schulze, 1958, 1961; Long, 1959; French, 1970; Mott, 1970). What they discovered was a tendency for the involvement of industrial leaders in community politics and affairs to change in both degree and orientation with the transitions to absentee ownership. The studies demonstrated that the leaders of locally owned industry commonly become deeply enmeshed in and, therefore, influenced by the social and political life of the community. Their involvement tends to be broad, direct, and based upon personal as well as organizational motives. Under absentee ownership this situation changes and the involvement of managerial leadership tends to become selectively restricted to primarily those issues and organizations that can have either a positive or negative impact on the political and economic interests of the operation as a satellite of its parent company.

Warren (1963) helps us understand this change in orientation by pointing to the significant change in reference groups that attends a transition to absentee ownership. Locally owned companies, in Warren's terms, are "horizontally integrated" into the community. In other words, for managers of locally owned companies, community leaders and citizens represent their significant others—the people upon whom their need for support, respect, and social position depend. Therefore, such managers and local entrepreneurs must give scrupulous consideration to the impact their business decisions have upon the welfare and opinions of the people with whom they share their community. For the local owner, company decisions are more than simply choices that have an impact on the economic success or failure of the company. They are decisions that can dramatically affect a wider range of social relationships for both oneself and one's family.

In contrast, the principal reference system for the top management of an absentee-owned company or plant is the corporation, and consequently the expectations and demands that come from corporate superiors must be given priority over any community considerations (Warren, 1963: 253). Thus, whether representing the company appropriately means staying out of community affairs entirely or aggressively defending corporate interests, the role of the executive of an absentee-owned operation must play the extremes of self-interest. To do otherwise might be career suicide.

The ultimate result of this shift in the alliances and orientation of a community's industrial leadership is the development of what Schulze (1958, 1961) has referred to as a "bifurcation of power." The unitary relationship that once existed among economic, social, and political roles is broken. The people who manage and control the major economic resources of the community become a group separate from the broader interests of the community as a place to live. Therefore, the people whose future and security are tied up in the community and who are most active in advancing and protecting its social and political life become the rest of the citizenry—schoolteachers, local labor leaders, small businesspeople, and so on. Herein lies the serious problem with a local economy becoming predominantly "vertically integrated" into other extracommunity systems as a result of absentee ownership.

With control over the economy having slipped from the community's hands, it is forced to depend upon the individuals and organizations whose concerns and interests hardly coincide with its own. From the standpoint of a parent corporation, a satellite operation is an investment, and the corporation's primary concern is with maximizing its return on that investment. The welfare of the community is at best marginal to the corporation's interest in maximizing its return on investment. Investment decisions are made in light of the entire corporate picture and not in terms of how many jobs they will save or create in the local community. Thus, although the upgrading of equipment and facilities, for example, might be essential to the continued viability of an operation as a profit-making enterprise, a parent company may choose not to pay for such improvements because a higher return can be obtained by investing in some other part of the corporation.

Obviously, local owners of manufacturing operations are not free to make this choice. If they wish to remain in business, they are forced to reinvest their earnings in the improvement of equipment and facilities. Similarly, simply to milk the business until it dries up could prove a great embarrassment to them and could also hurt many of the people with whom they will continue to live and associate on a daily basis. Such social sanctions simply do not exist for the absentee owner, and, therefore, a situation is produced in which the best interests of the local economy are not likely to be served.

This is particularly true in the industrialized cities of the Northeast, where the plants are often old, multistory buildings, the equipment outdated, the work force unionized, and the wages higher than in other parts of the country. To make such operations more profitable or capable of being profitable over the long haul necessarily requires considerable reinvestment of income into the operation. Much of the money that would have been available for making such improvements under local ownership is, however, lost to the management fee that a satellite plant must pay to its parent corporation. For these manufacturing operations to be upgraded, the corporation is required to make a substantial additional investment. From the view of the parent corporation, making this kind of an investment is likely to seem hardly worth it when it may be possible to build a new plant in the South or the Third World for the same or less money, and also avoid unions and high labor costs. Consequently, the northeastern manufacturing city is the loser. Similarly, because large corporations typically set standards for the average return they expect to receive from their various holdings, it

TABLE 5.1

Characteristic	Local Ownership	Absentee Ownership
Patterns of community power and decision making	unitary relationship between political and economic power	bifurcation of political and economic power
Pattern of integration of industrial enterprise in community	horizontal	vertical
Locus of control over industry	local	corporate headquarters
Economic significance of industrial firms to controlling management/ownership	major source of income	speculative—one of a number of investments
Social significance of firm performance and behavior to controlling management/ownership	central to individual and family status, prestige, and identity	peripheral
Dominant professional system for top management of local industrial firms	peers within community	corporate superiors
Motivational orientation of firm toward social and political community involvement	instrumental and affective	instrumental

is not unusual for a large company with many investments simply to close down a plant and go out of business, not because the operation fails to yield any profits, but rather because the profits are viewed as not high enough (Whyte, 1979: 15-16). Obviously, if an investment is not meeting corporate standards of profitability and requires major reinvestment to meet those standards, from the corporation's perspective it may simply make more sense to shut a firm down and obtain a sizable tax write-off.

As the foregoing discussion has attempted to illustrate, the industrial decline of the Northeast is, in part, a consequence of the increasing trend toward absentee ownership and loss of local economic control. Because of this trend, decisions regarding the maintenance and development of local firms that were once susceptible to local influences are now in the hands of organizations the central interests of which are far removed from those of the community. Table 5.1 attempts to portray these changing characteristics as communities shift from local to absentee control.

REACTING TO THE PROBLEM:
INDUSTRIAL WELFARE AND ECONOMIC WARFARE

In trying to cope with the predicament of lost existing industry and an inability to attract new industry, northeastern industrial cities might respond by developing a method to compensate for the control that was accrued to absent owners. This, in fact, is happening today. However, in attempting to exercise greater influence over the future of the community there has been a tendency to look to overly simplistic and even counterproductive measures as the answer. Notably, there has been an almost frantic grasping for tools at both the positive and negative extremes of the industrial incentive spectrum.

At the positive extreme, many northeastern industrial communities have turned to what verges on economic and political masochism in their efforts to repair the local economic base. They have basically taken the position, "If powers beyond the boundaries of our community are going to control the industrial base, then the best we can do is offer them every possible inducement to locate operations in our area." That is, like their southern competitors, northern cities have resorted to the use of long-term tax abatement and the doling out of low-interest industrial bonds as a means of attracting new industry. The major problem with the baiting of new industrial investors with virtually free land and services as well as low-interest money is that someone must ultimately pay for these "goodies" and, of course, that responsibility falls upon the shoulders of the average citizen and existing local business. Furthermore, these tactics obviously serve only to drive the local industrial base into deeper economic dependence upon forces external to the community with no added ability to influence the use of these resources.

More than sixty years ago, Thorstein Veblen (1923: 5-6) pointed to a growing awareness and discontent in the general population with the burdens imposed upon them by the increasing strength and predominance of absentee ownership of the economy:

> These time-worn principles of ownership and control, which are now coming to a head in a system of absentee ownership and control, are beginning to come in for an uneasy and reluctant reconsideration; particularly at the hands of the underlying population who have no absentee ownership to safeguard. Their questioning of these matters habitually takes the shape of personal recrimination and special pleading, but the drift of it all is no less evident. Those traditions that underlie the established order and that guide legal and administrative policy, proceed on an assumed community of class interests, a national solidarity of

interests, and an international conflict of interest; none of which is borne out by material facts. Therefore, these traditional policies that still govern the conduct of affairs, civil and political, legal and administrative, are also falling under suspicion of being incompetent, irrelevant and impertinent, if not downright mischievous. The losers under these rules of the game are beginning to see that law and politics, too, serve the needs of the absentee owners at the cost of the underlying population.

Today, this discontent has become aggressive political action in response to the many plant closures that have been enacted by absentee-owned corporations. It is in such action that we find the use of negative sanctions or disincentives to deal with and control the havoc that the decisions of absentee owners can have upon a local economy. Groups such as the Ohio Public Interest Campaign (OPIC), for example, have aggressively lobbied for legislation in their state that would require any firm employing more than 100 persons to announce shutdown plans at least two years prior to the anticipated date of closure, provide a severance payment of one week's wages for every year of service, and pay the community affected a sum equivalent to 10 percent of the operation's annual payroll (Cook, 1980: 38). Maine and Wisconsin have passed laws mandating 60-day advance notification before a shutdown. Similar bills are also on the books or under consideration in seventeen other state legislatures.

These efforts for legal control are the logical counterparts to the abatement tactics used to entice corporate investment. However, rather than trying to lure new investment, governments are attempting to coerce existing absentee-owned operations to stay in the community by imposing heavy economic sanctions if they choose to do otherwise. Though the social and economic justice of this approach, at least from the perspective of the average worker and the local community, may be greater than encouraging corporate irresponsibility through the use of industrial giveaway programs, this strategy by itself would seem to hold little promise as an approach for bringing renewed manufacturing strength to the troubled cities of the Northeast and Midwest. Surely, a new wave of corporate social responsibility and industrial expansion cannot be the expected return to states that have succeeded in holding industry an economic prisoner. Indeed, despite the costs, many companies will likely still close shop. Unless some alternative employment sources simultaneously emerge, both workers and their home communities will eventually be left without any solid economic grounding, though they will certainly be better compensated than they would have been without plant closing legislation.

In contrast to the two approaches outlined above, which alternatively rely upon either an "economic carrot" or an "economic stick" to achieve their ends, another group of grass-roots action strategies has also emerged during the past decade of efforts concerned with countering the problem of industrial decline. As the term "grass roots" suggests, the emphasis of these strategies is upon local choice and local initiative as opposed to looking primarily to larger, external economic and political forces to help the community solve its economic problems. That is, the focus of these approaches is on rebuilding and revitalizing the existing local economy through community-based cooperative action among labor, management, and local government. Federal and state assistance in such cases simply takes the form of support financing to help such efforts get off the ground (Whyte and McCall, 1980). Because the emphasis of these approaches is upon rejuvenating existing industry, they might best be termed strategies for "economic revitalization." We have been involved during the past ten years with conducting research into one significant revitalizing strategy that has developed primarily during the 1970s—area-based labor-management committees.

REJUVENATION THROUGH AREA LABOR-MANAGEMENT COOPERATION

One innovative strategy that has emerged in recent years and has proven itself to be a potentially powerful vehicle for revitalizing the local economy is the area labor-management committee (ALMC). The word "potentially" is used because not all the ALMCs formed in the past decade can really be considered serious efforts at community economic revitalization. Some have been little more than the old chamber of commerce "boosterism" of the past, cloaked with the participation of organized labor.

The first ALMC to be formed in the United States was the Toledo Labor Management-Citizens Committee in Toledo, Ohio. The Toledo committee was organized in 1945 as a result of the efforts of a local vice-mayor (Foltman, 1976). The reason for creating the ALMC vehicle was to try to bring industrial peace to the city that had developed a reputation for being a strike-prone town. At that time, the services of the Federal Mediation and Conciliation Service were not available and Ohio did not have a state mediation service. Therefore, it was felt that a local group was needed to intervene as a mediating force in instances in which labor and management could not successfully negotiate at the bargaining table by themselves. The Toledo ALMC was thus formed to provide this service. After hiring a committee director, a 48-member board was organized—composed of one-third labor, one-third management, and one-third public sector leadership—that has since mediated local contract disputes through the process described below:

> Initial mediation efforts are provided by the full time director. If he is unable to obtain a settlement, or if he deems it necessary, a tripartite panel of one member from the three sectors can be appointed by the chairman to mediate. The director and panel members call on other committee members for assistance [National Center for Productivity and the Quality of Working Life, 1978: 142].

Still an active and viable organization today, the Toledo Labor Management-Citizens Committee has been successful both as a mediating body and as a generally positive influence on the community's labor relations climate. It also became an impetus to other troubled communities, which, upon learning of the Toledo committee, used it as a model for organizing similar structures elsewhere.

The bulk of contemporary ALMCs emerged during the era of economic domination that began to afflict old, industrial U.S. cities in the early 1970s. One of these was developed in the city and surrounding region of Muskegon, Michigan (Woodworth, 1978). Situated on the shores of the Great Lakes and faced with a disappearing shipping industry, this city suffered from an increasing incidence of runaway plants, corporate buyouts of local entrepreneurs, strikes, and racial conflicts.

Eventually, in 1971 the Industrial Expansion Commission formed a team of outside consultants to plan Project Priority—with the mission of improving area labor relations, boosting regional productivity, and insuring job security. A board of three labor and three managerial representatives was formed to coordinate the work of various task forces in planning and carrying out programs consistent with Project Priority's objectives. By the mid-1970s, this innovative, collaborative, problem-solving process had solidified into a more formal structure, the Muskegon Area Labor-Management Committee (MALMC).

Perhaps the best-known ALMC in the United States, and one that has broken significant ground in themes of economic and social change, is the Jamestown Area Labor-Management Committee (JALMC) in western New York (Meek, 1981). Formed in 1972 under the leadership of the city's mayor, JALMC has become widely heralded for its successes and has served as a prototype for many ALMCs that have emerged in other areas of the United States.

Unlike the Toledo committee, the Jamestown committee is determined to maintain an arm's-length relationship with the collective bargaining process. By allowing contractual relationships between managers and unions to operate in traditional fashion, the JALMC has been able to emphasize issues on which both parties could collaborate— parallel activity to the bargaining process. By taking this direction, JALMC members have had to look for new ways to relate and work together.

The results have included some impressive accomplishments. One of the most interesting is that of multicompany and union involvement in designing skills-upgrading training programs to revive the community's depleted industrial skill base. Another is the participation of the entire workforce of one 300-employee operation in the process of planning a major plant redesign and expansion program. There have been substantial productivity gains in some companies as a result of JALMC stimulated labor-management cooperation, and new business has been brought into other operations as a result of joint labor-management problem-solving efforts.

WORKER/COMMUNITY OWNERSHIP
AND COMMUNITY RENEWAL

A second intriguing response to the problem of plant shutdowns that has emerged during the past decade is employee and/or community ownership. It is proving increasingly common for local labor and management groups and government officials to refuse to fatalistically accept notice from an absentee owner of the intent to close shop. Instead, local managers and workers, and, in some cases, even whole communities, have struggled together, pooling their resources in order to gain sufficient leverage with both private and governmental financial sources to literally buy out the operation from the absentee owners. More importantly, these new employee-owned and, in some cases, employee/community-owned firms are proving in a growing number of instances that it is actually possible for a local group to run and develop a firm more effectively and profitably than a large and affluent corporation (Stern and Hammer, 1978; Whyte, 1978; Zwerdling, 1980). Whyte (1978: 75) summarizes a few notably successful examples:

> When the Chicago and Northwestern Railroad became employee owned in 1972, it was considered to be in the weakest financial position of the Middle Western railroads. Since that time it has made a profit every year except in the recession year of 1975. Stock originally purchased for 83.5 cents is now valued between $10 and $11.

Vermont Asbestos Group became employee/community owned in 1974 with a stock then purchased at $50 per share. Recently, a private company offered $2,330 per share. . . .

In 1974, under conglomerate ownership, Saratoga Knitting Mill (Saratoga Springs, New York) lost money for the first time in its long history. In its first year under employee ownership, SKM expanded its work force from under 70 to over 120, set aside $1,000 per employee into an Employee Stock Ownership Trust, and declared a profit after taxes amounting to more than $1,000 per employee.

One recent conversion to worker ownership is the case of Jeanette Sheet Glass Corporation near Pittsburgh, Pennsylvania (Woodworth, 1980). Originally owned by ASG Corporation, the large 34-acre glass factory was closed in late 1978. The town of Jeanette was already suffering from the shutdown of a large tire company, and losing another 340 glass worker jobs (and a $6 million payroll) added to existing economic depression.

When only 10 percent of the employees secured other jobs, several labor leaders and a local attorney began to pursue the possibility of reopening the firm under local control. Union funds and industrial development grants were invested in a feasibility study to determine the likelihood of success in such a venture.

Eventually a $6 million funding package of loans, guarantees, and employee contributions was assembled and within a year the gates reopened. Workers became owners by each buying twenty shares of stock at $100 a share. Projected sales are between $15 and $16 million, a figure close to the amount of business being done before the firm was closed. New top managers were hired from outside the firm to bring in fresh ideas.

The initial months of 1980 at Jeanette suggested that a form of economic rebirth had occurred. Workers were excited about returning to their jobs, production was up, quality improved, and there was widespread optimism about the potential of this new experiment. The reality has included problems as well, the most serious being the decline of the building industry throughout the United States, which has backed up inventory at Jeanette. However, the skills and commitment of the company's worker/owners show a solid record after six years—a victory over tremendous odds.

In another case of absentee ownership closing down an operation and threatening the economic vitality of a community, Amsted Industries, Inc., determined in 1975 to shut its 70-year-old South Bend Lathe division. Although its profit picture was below the industry average, SBL was an important operation producing high-quality lathes, drills, and presses. The company's 500 employees were shocked that their $20 million business was about to fold.

Within a short time, the firm's president and a team of lawyers, bankers, and other technical assistance professionals had established an employee stock ownership trust based on a $5 million EDA loan to the city of South Bend, Indiana. The town passed it on to the worker's trust. An additional $2 million was loaned by banking interests, and SBL became worker owned with 10,000 shares of stock in the new trust.

In the first month of incorporation scrap was reduced by 70 percent, and the following months showed significant declines in absenteeism and turnover. Sales also improved as new orders increased 75 percent compared to the previous year. Productivity during the first twelve months resulted in a 25 percent increase. And the company was able to make its $750,000 repayment to the city seven months ahead of schedule (Ryan, 1976). As part of Amsted's conglomerate, SBL had sustained heavy losses its last five years. Now employee owned, its sales increased 53 percent and after tax profits grew to over $1 million, four times its profitability under Amsted.

Although the firm has retained hundreds of jobs and continued to contribute to the economic well-being of this midwest community, the past ten years have also included difficulties. An ongoing issue has been the fact that although employees have become owners, expectations that they would have more input into managerial decision making have not been met. Ownership on paper may be legal, but it does not guarantee control.

Another sticky issue has been the continuation of SBL as a unionized shop. Local 1722 of the United Steelworkers of America is not only made up of organized labor, but contains stockholders as well. Officials of the international union have been rather critical of the local union's actions in the move to worker-ownership. Dropping the pension and beginning the payment of stock dividends has resulted in considerable conflict. The tension between labor and management heightened to the point that the workers went on strike against the company they now own in the summer of 1980, a painful reminder of continuing dissatisfaction.

SEEDS OF RESISTANCE

Perhaps more significant than the singular success of a specific conversion to worker-ownership is the cumulative evidence that citizens, workers, and communities are fighting back at the arbitrary shutdown attempts of absentee owners. For instance, when distant owners in New York announced the closing of their 125-person rubber plant located approximately 2,500 miles away in Utah, a group of employees banded together to reopen the business as a worker-owned shop. In doing so they shook up city hall, the distant corporation, and the banks.

Similarly, when U.S. Steel declared its Ohio and McDonald Works in Youngstown, Ohio, would be shut, irate workers marched in protest at the corporation's headquarters in Pittsburgh. Eventually, members of United Steelworkers Local 1330 crashed the gate and took over Youngstown division offices. Chanting and singing, they unfurled banners from the rooftop of the administrative building, and held the facilities under their control until the corporation agreed to sit down and negotiate the possibility of the workers buying the factory and reopening it under joint ownership with local citizens as Community Steel, Inc. Although that effort was unsuccessful, a similar strategy is currently being used by steelworkers and public officials in Duquesne, Pennsylvania.

Rifkin (1977) reports the response of employees at the Stag Corporation's plant in Sorrel, Kansas, when informed by the executive offices in Chicago that production at the 35-year-old tractor assembly plant would be terminated. During the final two weeks of operation, representatives of the city council and the trade union made various proposals to salvage their life blood. The corporation was immovable and the whistle of the last shift blew. Instead of punching out and leaving the premises, workers remained on their jobs throughout the long night, working in shifts to allow for rest periods. Defying the order to terminate, this "wildcat production" continued for four days with townspeople bringing in food and needed supplies. Ultimately, company officials traveled to Kansas to meet with the city council and machinists' union. Agreeing on a settlement that would preserve jobs and the economic health of the community, Stag Corporation announced its willingness to sell the facilities to a jointly incorporated operation of the union and the city of Sorrel.

Thus the litany of responses to the threat of unemployment and community decline is becoming more aggressive and self-determining. Rather than accept welfare, move south, or take lesser paying or part-time work, people are countering by buying up closed firms. Employee ownership, though still in its infancy, seems to be growing. Latest estimates are that between six and seven thousand such firms currently exist in the United States. They range from small enterprises such as the Atlas chain in Pennsylvania, which is equally owned and managed democratically by several dozen workers, to large worker-owned companies such as Eastern Airlines, with 30,000 employees.

Some, such as the sixteen plywood firms in Washington and Oregon, are highly participative, characterized by consensual decision making, workers on the board of directors, and members holding equal shares of stock and voting rights. Others, such as the Okonite Company in Ramsey, New Jersey, were restructured into an ESOP mainly to become free from outside corporate control. However, Okonite employee-owners cannot vote, have no participation on the board, and the company does not even share basic financial information (Ross, 1980).

The range of industries and products that have experimented with employee ownership are also varied. In addition to the businesses discussed here, there are publishing firms such as the Milwaukee *Journal* and Boston's *Real Paper*, rubber products such as Republic Hose Manufacturing in Ohio, food industries such as Sea-Pack Corporation in Georgia and Texas, and garment and textile firms such as Saratoga Knitting Mill in New York and Bates Fabrics, Inc., in Lewiston, Maine. There are steel plants of blue-collar employees such as Weirton Steel in West Virginia, and modern offices of white-collar insurance workers such as Consumers United Group (CUG) in the heart of Washington, D.C.

CONCLUSION

The central thrust of this article has focused on the unraveled social and economic fabric of communities victimized by absentee ownership. The concomitant pain of impersonal business decisions taken with no sense of local social responsiblity, runaway industry, growing unemployment, and loss of community tax base all combine in a conspiracy of neglect wrought by distant owners and executives.

In attempting to cope with this bifurcation of power and self-interest, the past decade has been replete with community/worker tactics to create alternatives beyond the assumption that industrial shutdowns are inevitable and irreversible. The emergence of area labor-management committees as a vehicle for problem-solving and economic revitalization seems to show significant potential. At a more fundamental level of change, the shift to employee ownership and worker/community control suggests the possibility of reintegrating political and economic dimensions of society into a more holistic and democratic system.

Although such developments may be too new to evaluate fully, initial evidence leans toward the implication that these mechanisms may result in more productive and effective institutions that yield a higher-order quality of life for all concerned.

REFERENCES

AHERN, R. W. (1979) The Area-Wide Labor Management Committee: The Buffalo Experience. Buffalo: Buffalo-Erie County Labor-Management Council.

BECKMAN, J. W. (1974) Industry in New York: A Time of Transition. Legislative Document No. 12. New York: New York State Legislature.

CHINITZ, B. (1977) "Manufacturing employment in New York State: the anatomy of decline," pp. 57-94 in B. Chinitz, The Decline of New York in the 1970s. Binghamton, NY: Center of Social Analysis, SUNY.

CLARK, T. N. (1967) "Power and community structure: who governs, where and when?" Sociological Quarterly 8: 291-316.

CLELLAND, D. A. and W. H. FORM (1964) "Economic dominants and community power: a comparative analysis." American Journal of Sociology 59: 511-521.

COOK, D. D. (1980) "Laws to curb plant closings." Industry Week 204: 26-41.

DAHL, R. A. (1961) Who Governs. New Haven, CT: Yale University Press.

FOLTMAN, F. F. (1976) Labor-Management Cooperation at the Community Level. Ithaca: Cornell University, New York State School of Industrial and Labor Relations.

———and P. D. McCLELLAND (1977a) New York State's Economic Crisis: Jobs, Income and Economic Growth. Ithaca: Cornell University, New York State School of Industrial and Labor Relations.

———(1977b) "Introduction and summary: the extent of the problem," pp. 1-16 in F. F. Foltman and P. D. McClelland (eds.) Economic Crisis. Ithaca: Cornell University, New York State School of Industrial and Labor Relations.

FOLTMAN, F. F., J. H. SRB, and M. BLONDMAN (1975) Economic and Job Trends in New York State. Ithaca: Cornell University, New York State School of Industrial and Labor Relations.

FRENCH, R. F. (1970) "Economic change and community power structure: transition in cornucopia," pp. 181-192 in P. E. Mott and M. Aiken (eds.) The Structure of Community Power. New York: Random House.

HUNTER, F. (1953) Community Power Structures. Chapel Hill: University of North Carolina Press.

JANOWITZ, M. [ed.] (1961) Community Political Systems. New York: Free Press.

LINDSAY, R. (1975) "The decline of manufacturing in New York State," pp. 7-28 in F. F. Foltman et al. (eds.) Economic Trends in New York State. Ithaca: Cornell University, New York State School of Industrial and Labor Relations.

LONG, N. E. (1959) "The corporation, its satellites, and the local community," pp. 202-217 in E. S. Mason (ed.) The Corporation in Modern Society. Cambridge, MA: Harvard University Press.

LYND, R. S. and H. M. LYND (1937) Middletown in Transition. New York: Harcourt Brace Jovanovich.

MASON, E. S. [ed.] (1959) The Corporation in Modern Society. Cambridge, MA: Harvard University Press.

MEEK, C. (1981) "Labor/management cooperation and economic revitalization: the story of the growth and development of the Jamestown Area Labor-Management Committee." Ph.D. dissertation, Cornell University.

MOTT, P. E. (1970) "The role of the absentee-owned corporation in the changing community," pp. 170-180 in P. E. Mott and M. Aiken (eds.) The Structure of Community Power. New York: Random House.

———and M. AIKEN [eds.] (1970) The Structure of Community Power. New York: Random House.

National Center for Productivity and the Quality of Working Life (1978) Directory of Labor-Management Committees. Washington, DC: Government Printing Office.

O'LEARY, J. (1974) Industrial Development in New York State, Albany: Legislative Commission Expenditure Review.

PELLEGRIN, R. J. and C. H. COATES (1956) "Absentee-owned corporations and community power structure." American Journal of Sociology 11: 413-419.

RIFKIN, J. (1977) Own Your Own Job. New York: Bantam.

ROSS, I. (1980) "What happens when the employees buy the company." Fortune (June 2): 108-111.

RYAN, J. J. (1976) "How and why U.S. helped some workers take over a machine-tool manufacturer." Wall Street Journal (August 16).

SAYRE, W. S. and H. B. KAUFMAN (1960) Governing New York City. New York: Russell Sage Foundation.

SCHULZE, R. O. (1961) "The bifurcation of power in a satellite city," pp. 19-80 in M. Janowitz (ed.) Community Political Systems. New York: Free Press.

———(1958) "The role of economic dominants in community power structure." American Sociological Review 23: 3-9.

SMITH, P. A. (1965) "The games of community politics." Midwest Journal of Political Science 9: 37-60.

SRB, J. H. (1975) Conclusions and Recommendations of the Labor-Management Conference on the Business Climate and Jobs in New York State. Ithaca: Cornell University, New York State School of Industrial and Labor Relations.

STERNS, R. N. and T. H. HAMMER (1978) "Buying your job: factors affecting the success or failure of employee acquisitions." Human Relations 31: 1101-1117.

THERNSTROM, S. (1965) " 'Yankee City' revisited." American Sociological Review 30: 234-242.

VEBLEN, T. (1923) Absentee Ownership and Business Enterprise in Recent Times: The Case of America. New York: Sentry. (Reprinted, 1964)

VIDICH, A. J. and J. BENSMAN (1958) Small Town in Mass Society. Princeton, NJ: Princeton University Press.

WARNER, W. O. and J. O. LOW (1947) The Social System of the Modern Factory. New Haven, CT: Yale University Press.

WARREN, R. L. (1963) The Community in America. Chicago: Rand McNally.

WHYTE, W. F. (1979) "Confronting the conglomerate merger menace." Statement to the Committee on Small Business Subscommitee on Anti-Trust and Restraint of Trade, House of Representatives, U.S. 96th Congress, 2nd Session.

———(1978) "In support of employee ownership." Society 15: 73-82.

———and D. McCALL (1980) "Self-help economics." Society 17: 22-28.

WILDAVSKY, A. B. (1964) Leadership in a Small Town. Totowa, NJ: Bedminster.

WOODWORTH, W. (1980) "A counter to plant shutdowns." Self-Management 7: 29-32.

———(1978) "Consulting with conflicting parties: a method for achieving mixed results," pp. 147-152 in J. C. Susbauer (ed.) Academy of Management 1978: Proceedings. Wichita, KS: Academy of Management.

ZWERDLING, D. (1980) Workplace Democracy. New York: Harper & Row.

PART II

REVITALIZATION THROUGH LABOR-MANAGEMENT COOPERATION

Historically, cooperative initiatives by unions and managers have arisen at critical periods in the United States as attempts to resolve serious disputes and minimize the negative consequences of the adversarial relationship between the two parties. During World Wars I and II, management and labor worked together on the National War Labor Board not only to ensure industrial peace but also to improve productivity. Cooperative relationships were also structured years ago in the rail industry as unions and management attempted to cope with economic stress by reducing grievances, improving working conditions, and altering work rules and procedures.

Although perhaps over 5000 labor/management committees have existed in the past, the movement eventually diminished with the end of World War II. However, a broader, more community-based strategy soon developed, first in Toledo, Ohio, and then in Louisville, Kentucky. Known as Area Labor-Management Committees (ALMCs), the two parties were joined by local governments to referee interaction and resolve disputes. Thus a tripartite system at the regional level began to resolve new problems and explore local needs. These efforts, in contrast to earlier systems, were introduced not to supplement the institution of collective bargaining, but to complement it. Where needed, and at appropriate times, traditional elements spelled out in the labor contracts were adhered to. But in other circumstances, ALMCs extended the relationship of the two parties into new dimensions of mutual collaboration.

The chapters in Part II analyze organizational processes and economic outcomes that may occur as part of the creation of regional labor/management committees. More specifically, the authors introduce conceptual and practional realities involved in the recent expansion of ALMCs. In Chapter 6, Joel Cutcher-Gershenfeld pulls together a number of issues that have grown out of his experience with the Michigan Quality of Working Life Council. Forces leading to the formation of this adjunct approach to industrial relations are analyzed within the context of communities and local realities. Forces that initially gave rise to ALMCs are juxtaposed against oppositional forces that may impede cooperative labor-management problem solving.

In Chapter 7, Warner Woodworth describes in detail the process of achieving joint labor-management action in Muskegon, Michigan. It is an important case of a special form of third-party assistance to management and labor in the start-up phase of an ALMC. The sequence of consultation activities, as articulated by one of the community's professional resources, is spelled out in detail. The struggles and setbacks are highlighted by the ALMC's attempts to reverse decades of economic disintegration and labor-management hostilities. The results of this grass-roots effort at regional revitalization have important implications for other groups just beginning such a process.

Christopher Meek, in Chapter 8, reports on the important development of the Jamestown, New York, ALMC. He describes Jamestown's unique structure and spectacular results based on five years of on-site field research. The community's response to a growing economic crisis is reconstructed, revealing the expansion from a micro-effort into a macro-level approach to area revitalization. In contrast to a number of other cases, the interventionist role of a progressive mayor in Jamestown appears to be an unusual factor in the rise of the ALMC. Meek concludes by summarizing a number of the impressive economic results achieved as this old industrial town's reputation is transformed, hundreds of jobs are saved, plant closings are blocked, and more productive enterprises are created.

The final chapter in Part II, by Robert Ahern of the Buffalo-Erie Labor-Management Council, is important because of Ahern's key role in ALMC field experience. The Buffalo case provides useful insight into the complexities of labor-management cooperation in the big city, and Ahern has a keen sense of the steps involved in establishing an ALMC. Job training programs, dispute resolutions, and the concept of satellite ALMCs were instrumental in achieving the impressive performance results in Buffalo.

6

THE EMERGENCE OF COMMUNITY
LABOR-MANAGEMENT COOPERATION

Joel Cutcher-Gershenfeld

Recent decades have seen the emergence in the United States of a new feature inhabiting both the industrial relations landscape and the terrain of economic revitalization. Labor-management cooperation at a community level is born of both worlds and has the potential to have a significant impact on each. For most observers the critical question is, Will this potential be realized and, if so, what are the implications?

Conventional wisdom holds that labor and management will only step out of their traditional roles and work together cooperatively in the face of a crisis. Although there have been many crises that have not prompted cooperation, there are indeed plenty of instances in which a combination of critical events has brought together these traditional adversaries. However, the emergence of such cooperation is generally accompanied by a curious assumption. It is assumed that joint initiatives will endure only so long as the crisis persists. In fact, the evidence indicates there are many factors other than crises that determine whether such initiatives will endure and succeed.

TABLE 6.1
Area Labor-Management Committees in the United States

Year Established	Committee
1945	Toledo, Ohio
1946	Louisville, Kentucky
1953	Chattanooga, Tennessee
1958	Jackson, Michigan
1963	South Bend, Indiana
1965	Green Bay, Wisconsin
1969	Western Kentucky (Paducah)
1970	Fox Cities, Wisconsin
	Upper Peninsula, Michigan
1972	Jamestown, New York
	Pittsburgh, Pennsylvania
1975	Buffalo/Erie County, New York
	Chautauqua County, New York
	Clinton County (Lock Haven), Pennsylvania
	Cumberland, Maryland
	Evansville, Indiana
	Mahoning Valley (Youngstown), Ohio
1976	Elmira, New York
	Springfield, Ohio
1977	Lansing, Michigan
	Muskegon, Michigan
	North Central, Wisconsin
	Riverside/San Bernardino, California
	St. Louis, Missouri
	West Virginia
1979	Akron, Ohio
	Beaumont, Texas
	Duluth, Minnesota
	Haverhille, Massachusetts
	Portsmouth, Ohio
	Scranton, Pennsylvania
	Sioux City, Iowa
1980	Elk County, Pennsylvania
1981	Kankake, Illinois
	Kenosha, Wisconsin
	Philadelphia, Pennsylvania
1982	Chillicothe (Ross County), Ohio
	Danville, Illinois
	Decatur, Illinois
	Manistee, Michigan
	Norwalk (Fireland), Ohio
1983	Aurora, Illinois
	Columbus, Ohio
	Des Moines, Iowa
	Erie, Pennsylvania
	Memphis, Tennessee
	Mercer County, Pennsylvania
	Minneapolis, Minnesota
	New Castle, Pennsylvania
1984	Oakland County, Michigan

SOURCE: Updated from Cutcher-Gershenfeld (1984: Table 1).
NOTE: Although the vast majority of these committees are still active, a few are no longer in operation or have departed from their labor-management focus.

In this chapter, these factors will be identified and analyzed in the course of tracing the emergence of area labor-management committees (ALMCs) in a variety of U.S. communities.[1] Over forty of these organizations have been established since the end of World War II (see Table 6.1). Although they have been and remain somewhat fragile as organizations, we will see that they have proved an important source of innovation.

ALMC STRUCTURE
AND OPERATION

Most writing on ALMCs has been limited either to a catalogue of activities (for example, see Siegel and Weinberg, 1982: 25-97) or to a linear, developmental mode (e.g., conception, gestation, birth, maturation, death; see Leone et al., 1982). In fact, the establishment and operation of an ALMC is a dynamic process, influenced by many factors. A central purpose of this chapter is to identify such factors and conduct a preliminary assessment of their explanatory and predictive power. It is only in the context of this sort of dynamic model that it is possible to consider the implications of ALMCs for industrial relations or economic revitalization.

In examining any particular ALMC, it is important to consider at least the following set of factors:

(1) the forces that led to its formation
(2) the particular characteristics of the community (size, industry mix, and so on)
(3) the particular characteristics of the ALMC's leaders and any third-party neutrals
(4) the evolving needs of labor and management in their primary industrial relations roles
(5) the emerging needs of the ALMC as an organization
(6) the nature and strength of forces opposed to the functioning of the ALMC
(7) the forces beyond the community, such as changes in state and federal public policy or national shifts in the economy, the composition of the work force, and the practice of industrial relations

Although each factor will be examined in this section of the article, we will see that they are all closely intertwined as explanations of why ALMCs do what they do.

FORMATIVE FORCES

The initial forces or events that led labor and management leaders to step out of their traditional roles are the focus of most discussions about labor-management cooperation. They are important in that it is a hard step, for which the parties' traditional roles provide little preparation. The parties' principal modes of interaction are through collective bargaining and the grievance procedure. However, collective bargaining is best suited for discrete (rather than continuous) issues (that is, issues that can be successfully addressed once every two or three years). At the same time, grievance procedures have evolved into relatively legalistic rule enforcement systems (Kuhn, 1967: 252). Thus with a few exceptions, the core institutions of industrial relations are not oriented toward continuous problem solving. Additionally, as Thomas Kochan (1980: 415) observed, "There is tremendous inertia in the [collective bargaining] system."

The importance of initial forces in bringing parties together was recently documented in a U.S. Department of Labor study of joint initiatives in eight U.S. communities. This study found that all the communities were facing a long-term crisis rooted in declining employment opportunities and compounded by deteriorating labor-management relations (Leone et al., 1982: 67). In most cases, the labor and management leaders were galvanized into action by a specific event or combination of events, such as a particularly long and bitter strike or a major plant closing.

Social theorist Eric Trist (1978) observes that such critical situations often reveal a larger societal problem[2] that has not been addressed adequately by existing social institutions. There are a handful of situations in which labor and management leaders have come together to address the larger set of social and economic problems before a critical situation develops. However, in the majority of cases it takes a crisis to bring the parties together. This is not surprising; the role of crisis pressures in unfreezing established social relationships is well documented in a variety of disciplines. In particular, Kurt Lewin's (1947) three-stage model of social change highlights the importance of initial "unfreezing" forces.

To an extent, these initial forces are subsequently reflected in the structure and operation of the committee that is established. Thus ALMCs were formed in Toledo and Louisville[3] shortly after World War II in response to a postwar wave of strikes. These were tripartite organizations, the principal purpose of which was to help facilitate the resolution of collective bargaining disputes. As a result, they chose to function with relatively little staff and found it appropriate to adopt a quasi-official role in city government.

Many ALMCs faced with a crisis linked to a plant closing or economic development were similarly informal at this early stage. But their focus was industrial retention. Thus in Jamestown, the initial task confronting the organization was the potential closing of at least a half-dozen local firms. The resolution involved the informal opening of lines of communications and a series of joint efforts to generate capital or other needed resources.

Similarly, labor and management leaders came together in Des Moines, when intergovernmental squabbling over federal sewer system monies threatened the loss of this critical part of local infrastructure. Here, too, the initial structure was informal and task-force oriented in nature. Most of these emerging committees chose to be separate from local government or other existing institutions. At this early stage, the crisis-driven ALMCs are able to operate with a relatively limited set of implicit or explicit ground rules. Aside from the early mediation committees, these have typically included agreement to the following:

(1) Participation in the ALMC will be voluntary.
(2) Representation from labor and management will be equal.
(3) Interference in collective bargaining and grievance handling will be prohibited.

Sometimes the following two ground rules are also present:

(4) All major decisions will be made by consensus.
(5) The activities of the ALMC should not result in the loss of jobs (Gershenfeld and Costanzo, 1982: 11).

Interestingly, ALMCs that are not formed in the face of an immediate crisis tend to take on a more formal structure at the outset. This was the case in Lansing, where the parties came together primarily out of a long-term interest in community education and assistance on the subject of labor-management cooperation—especially in the area of improving the quality of work life (there was also some concern about a few instances of industrial exodus). Early on, this organization not only had ground rules such as those identified earlier, but had formed bylaws and an explicit program of courses, conferences, and technical assistance.[4]

Although crises have certainly played critical roles in bringing together labor and management leaders in various communities and have even left some imprint on the structure of ALMCs, factors other than the specific crisis are central to understanding the formal and broadly focused programs that many ALMCs have established. In the context of Lewin's model for social change, the evolution and subsequent effectiveness of these organizations can best be understood by looking beyond "unfreezing" forces. Instead, it is best to look at forces influencing the middle and final stages of the model, which consist of experimentation with new social relationships (Stage II) and institutionalization of one or more of the new patterns (Stage III). With respect to ALMCs, the six factors discussed below are the most salient.

COMMUNITY CHARACTERISTICS

The very nature of a community (especially its work force and industry mix) sets important parameters on the activity of an ALMC. In a large community, for example, labor and management leaders play a very different role from that they perform in a small community. In the former, they are but one of many interest groups. As a result, their common bond is, in part, to serve as a coalition that is stronger than either would be if they were separate. Thus labor and management addressed the sewer issue in Des Moines or lobbied for a unique approach to job training in Buffalo. In Philadelphia the parties not only pursued dredging the Delaware River as a shared concern, they also explicitly identified influencing public policy as an objective of their organization.

ALMCs in smaller communities may have just as strong an influence on public policy, but they rarely articulate such as a goal. This probably stems from the great overlap that occurs among the roles of ALMC leaders. In Cumberland, the management cochair of the ALMC was recently elected as the city's mayor and many other principals from both sides of the table hold high elected and appointed offices. Similarly, in Manistee, Michigan, membership in the recently formed ALMC includes key members of both the city and county councils. In Jamestown, the ALMC's coordinator is also the city ombudsman. In Muskegon, the ALMC's executive director is also head of the local private industry council (PIC). Similar links exist for most active, small-community ALMCs.

The size of the community not only influences whether public policy is an ALMC priority, but affects the choices made by the committee to provide certain services. Thus in the area of job training the Philadelphia ALMC is but one of many successful bidders for private and public funds to provide a specific service—the training of displaced workers. By contrast, the Jamestown ALMC, in conjunction with the local community college, coordinates a vast array of skills training programs that might not otherwise be available in the community.

Naturally, the local industry mix and work force conditions are other critical factors shaping the ALMC. In Beaumont, construction constitutes a major portion of the unionized work force and cooperation with that industry has been the principal focus of community initiatives in that area. In Jamestown, the skills training program first emerged when it was discovered that the average age in the local metal- and wood-working industries was 55 and that workers were retiring without passing along their skills. In Manistee, the presence of a high proportion of facilities owned by outside organizations was a source of frustration because local management was either unable to make key commitments or at least constrained regarding how quickly it could make promises. Also, the existence of Caterpillar, Inc., as a single dominant employer in Aurora, Illinois, raised a complex series of fears on the part of other employers and unions during discussions about how to structure an ALMC.

Finally, an important local characteristic is the degree of unionization among the local force. The 1982 U.S. Department of Labor study of eight ALMCs found that all had substantially higher than average degrees of union penetration (Leone et al., 1982: 41). In some ALMC communities, such as Manistee, over 75 percent of the work force is organized. Not surprisingly, there have been few efforts to establish ALMCs in communities with relatively small unionized sectors. Indeed, in Shiawasse County, Michigan—where a number of workers were organized—efforts to establish an ALMC were still hampered because the largest local employer operated on a nonunion basis. Considering the power and influence an ALMC can acquire in a community, it is not surprising that management would typically choose to join with labor only where it spoke for a significant proportion of the work force. Conversely, where labor is in a dominant position in a community, it is likely to perceive cooperation with management as a dilution of its power.

The community most likely to support an ALMC, then, is one in which unions represent a significant portion of the work force and power relations between labor and management are relatively balanced. Within that context, the size of the community, the local industry mix, and the characteristics of the work force will set boundaries on the actual scope and direction of the ALMC.

LEADER AND NEUTRAL CHARACTERISTICS

For every community with an ALMC, there are others that are comparable in every respect, but that do not support such an organization. As one researcher has observed, "The existence of a labor relations and/or community economic development crisis is necessary [but] not sufficient for committee development" (Blondman, 1978: 125). In fact, Blondman's analysis went so far as to identify the existence of leaders of foresight and dedicated third-party individuals skilled in facilitator roles as the most critical factors in explaining the creation of an ALMC. Blondman's consequent use of entrepreneurial theory to analyze an ALMC is a valuable contribution to the literature, but the analysis should be tempered by the central thesis of this chapter—which is the dynamic interplay of *many* factors in shaping an ALMC.

During the formation of an ALMC one task for labor and management leaders is the articulation of fundamental concerns. This is to provide both the image and reality that the leaders are not abdicating their primary allegiance. Thus when the possibility of establishing an ALMC was first discussed in western Kentucky, Jamestown, and other places, labor leaders initially charged that the business community had no interest in cooperation—that it really wanted to keep new industry out in order to keep wages low. Typically, business countered that no one would want to move into the community given the actions of its existing unions. At the outset, an ALMC is only possible where leaders are willing to, if you will, put their cards on the table.

Of course, the combination of leaders and circumstances that make an ALMC possible involves more than just the articulation of concerns. It also involves being able to move beyond specific concerns and explore potential shared goals. It is here that real leaders emerge. It is also here that key assistance has often come from a broad array of third parties, including federal mediators (for example, in Jamestown, Evansville, Clinton County, Cumberland, South Bend, Upper Peninsula); local citizens—attorneys, clergy, and others (for example, in Jamestown, Cumberland, Jackson, Scranton); state government officials (for example, in Manistee); individuals from colleges or universities (for example, in Lansing, Aurora, Des Moines, Muskegon); and statewide, regional, or national labor-management organizations (for example, Manistee, Haverhill, Erie). Indeed, with the possible exceptions of Buffalo and Pittsburgh, third-parties have played central roles in the establishment of all ALMCs.

An important recent trend is the growing network of individuals who are specifically reaching out to communicate where there is interest in establishing an ALMC. This network includes (1) representatives of the National Association of Area Labor-Management Committees (NA-ALMC), (2) many federal mediators, (3) an innovative outreach program established at the Michigan Quality of Worklife Council, and now incorporated into the objectives of a statewide ALMC network, (4) past ALMC staff that move to new communities, and (5) university professors engaged in "action research." Although it is clear that organizing an ALMC is more complex than most forms of community organizing, a set of common third-party approaches is emerging.[6]

Of course, leaders and third parties are important, not just to make an ALMC possible, but also because their personal characteristics are continually reflected in the organization as it forms and evolves. Thus in a community such as Jackson, the members and neutrals have offered informational seminars on work-site quality of work life (QWL) initiatives, but are generally opposed to playing an active role in assisting parties to set up such efforts. By contrast, nearby Lansing boasts an ALMC in which various forms of QWL assistance are perhaps its most important goal. In part, this difference stems from the contrasting reasons for the formation of these ALMCs and from the different mix of unions and employers they each contain. (GM, Motor Wheel, UAW, CWA, and other organizations that were prominent in Lansing are noted for their support of QWL.) However, these differences also reflect a conscious choice on the part of leaders and neutrals.

One additional implication of the involvement of various leaders and neutrals occurs with respect to public-sector labor relations. For ALMCs that are formed to help turn around a contentious labor-management climate, the concern is generally with private-sector relations. For this reason, and because of a concern for maintaining a nonpartisan posture, many ALMCs have initially defined their focus as limited to the private sector. However, ALMCs in Jamestown, Kenosha, Lansing, Evansville, Cumberland, and elsewhere have discovered great interest in work-site labor-management committees on the part of labor and management in schools, hospitals, social service organizations, police, fire, and other public agencies. The consequence of such activity is that it transforms various "neutral" public officials into employers and expands the range of potential union members of the organization.

Finally, although it should be obvious, an ALMC is also shaped by the ranks of labor union members and the host of middle- and lower-level managers in a community. These individuals attend ALMC events, benefit from ALMC-sponsored training, donate time to help with major public events, and participate in various work-site joint initiatives. Most important of all, on the labor side, they choose whether or not to reelect leaders who have taken a strong stance for (or against) the ALMC.

INDUSTRIAL RELATIONS NEEDS
OF THE PARTIES

No matter how highly committed a community's labor and management leaders are to an ALMC, their primary responsibilities will always remain in the industrial relations arena. Indeed, Kochan and Dyer (1976: 68) argue that involvement in a joint committee will only be forthcoming when the parties see it as "instrumental to the attainment of the goals valued by their respective organizations." Various pressures and political considerations within companies and unions all have the potential to constrain ALMC activities.

This helps explain the highly pragmatic or instrumental nature of most ALMCs—they focus on activities such as technical education, training, and consulting assistance, which are seen as directly useful by a union membership or a local employer. It also helps explain the pressure that labor and management leaders experience when, as is often the case, it takes far longer than anticipated to establish ALMC programs. It even helps explain why the majority of ALMCs (with a few notable exceptions) have ground rules prohibiting interference in collective bargaining (because that might interfere with primary goal attainment).

This close link between industrial relations roles and ALMC endurance and evolution is consistent with the historical experience of other labor-management committees. Though many of these other committees have been formed in the face of a crisis, their ultimate effectiveness has had more to do with their relationship to the parties' core, institutional interests, which traditionally have been reflected in collective bargaining. Initiatives established as an *alternative* to collective bargaining (ranging from the 1910 Protocol of Peace to WWI and WWII shop committees) have failed as soon as at least one of the parties had the chance to choose between the alternative and collective bargaining. By contrast, those serving as a *complement* to collective bargaining (ranging from the 1900 National Civic Federation to railway and textile committees in the 1930s to today's construction committees and health and safety committees) have endured—so long as they were effective in that complementary role.[7]

However, Kochan and Dyer's argument, which was presented in 1976, and the central premise that industrial relations roles are primary, must be updated in one important respect. They assume an industrial relations system that is relatively stable. Yet, as we are beginning to discover, the 1970s was an era of mounting tension in the U.S. industrial relations system. Now Kochan and Piore (1984) point out that these pressures at the levels of public, corporate, and union policy; collective bargaining; and workplace practices have driven the system into a state of major transition. Factors as diverse as shifting world markets, an increasingly white-collar work force, decreasing union influence in public policy, corporate union-free and concession strategies, increasingly decentralized bargaining, a scope of bargaining expanding into "continuous" issues, computer technologies, and increasing worker-participation in decision making all conspire to ensure that traditional notions of industrial relations are no longer sufficient.

ALMCs (at least the majority that have been established in the 1970s and 1980s) are clearly among the offspring of this transition. They are born of a realization that an unguided industrial relations system can exacerbate corporate disinvestment, undermine local industry, and prevent both the joint exploration of new technologies and the joint redesign and reorganization of work. Thus, although most ALMCs have policies against interfering directly with collective bargaining, one of their central goals is to improve the quality of labor-management relations.

This leads most ALMCs to sponsor joint seminars, conferences, and other public events not just for their content, but to provide labor and management with the chance to interact outside their adversarial roles. Opportunities also occur as the parties work together to establish specific programs for skills training, industrial expansion, and the like. Further, beginning with the Jamestown ALMC, many of these organizations are pioneering the unique notion of providing community-based assistance to local unions and employers seeking to establish work-site joint committees. Sometimes these are just top-level committees that engage directly in problem solving and sometimes they are intended as steering committees for a variety of shop-floor and middle-level problem-solving committees or groups.

With respect to the work-site committees, there is a growing body of anecdotal and some case study evidence of top-level joint committees or shop-floor committees that have helped increase communications and respect between labor and management, improve production processes and product quality, promote corporate investment, improve working conditions, and shift the flow of information and decisions in organizations. Although attempts to measure the impact of such participative activities formally are inconclusive (see Katz et al., 1983: 14), there is well-documented research indicating that overall industrial relations performance does have a significant impact on plant performance (Katz et al., 1983: 9; Ichniowski, 1984: 28). The task that remains is to formally measure the impact of ALMC activities on industrial relations performance. Still, all the indirect activities noted in the preceding paragraph and the direct activities noted below do offer indications that at least some ALMCs do have a positive effect on work-site industrial relations performance.

The ALMCs in Toledo, Louisville, and Buffalo (to a lesser extent) have formally indicated a willingness to assist in the resolution of collective bargaining disputes. But such assistance even occurs in the case of an

ALMC that has a formal policy against interference in collective
bargaining disputes. It is via the emerging network of relationships that
most ALMCs lead to a surprisingly high degree of informal peer
assistance (some would say peer pressure) in the resolution of disputes.

The outcome of this activity goes beyond meeting the parties'
industrial relations goals—it changes the industrial relations system in
ways that may make it more viable. For both parties, it offers a path out
of a cycle in which, as one Manistee manager put it, "You make your
final offer, there's a strike, and then you make your final, final offer." A
strike rate far above the national average has been observed in many
communities prior to the formation of ALMCs (Leone et al., 1982: 46)
and that history is merely suggestive of an adversarial heritage that is
sometimes generations old. After the establishment of ALMCs in
Manistee, Jamestown, Cumberland, Buffalo, Evansville, and many
other communities, the strike rate has not only plummeted, but a variety
of early innovative settlements have emerged.

The scope of this change was evident, for example, when the many
agreements that came up for bargaining in the first year after the
establishment of an ALMC in Manistee were all negotiated without a
strike, although in previous years, negotiations rarely occurred *without*
a strike. In Cumberland, the parties involved in one agreement were able
to break out of a cycle whereby inventories were always built up prior to
bargaining, leading to either massive layoffs (if a settlement was reached
quickly) or a strike. They agreed on early negotiations that would
culminate many months prior to the scheduled contract expiration,
offering the incentive that any negotiated increase in wages would be
effective immediately (rather than at expiration) and the caveat that
they would begin at "ground zero" if the early effort did not succeed.
Federal mediators—a useful barometer of local labor-management
climates—report changes not only in outcomes (both the number of
strikes and the content of agreements), but in the way the parties choose
to interact.[8] This sort of self-monitoring is far preferable to efforts to
mandate industrial relations behavior. It also introduces to the system
the critical capability for improving itself.[9]

For union leaders an additional and essential outcome of being
involved in a very active ALMC is increased community recogniton for
labor's distinctive role and competence. This was reflected symbolically
when the Muskegon Chamber of Commerce arranged for former UAW
president, Douglas Fraser, to be the keynote speaker at its annual
banquet, or when business leaders marched in the Maristee Labor Day
parade. It is also evident through concrete increases in labor's power and
influence, such as the election and appointment to public office of labor
leaders active in ALMCs in Cumberland, Manistee, and elsewhere.

John Popular, executive director of the National Association of ALMCs, observes that the role of labor unions in many communities is that of the outsider looking in on decision making or that of a separate interest group typically opposed to business. By contrast, he notes, an ALMC can provide local unions with access to the community power structure and acceptability (cited in Cutcher-Gershenfeld, 1984: 82). Though not traditionally a central goal of the labor movement, it is increasingly clear that such influence will be a priority in the emerging industrial relations system.

For management, the ALMC model holds the potential for a shift in workplace industrial relations. In big cities, such as Buffalo, in small cities, such as Jamestown, and in many other areas, joint work-site initiatives have led to cooperation on matters ranging from bidding on new business, improving quality control, developing effective gain-sharing programs, and improving the work flow, to even redesigning the facility.

Nationally, there are certainly those in labor and management who would take issue with a system that involves increased political power for labor, increases flexibility for management, and encourages a higher community profile (formally or informally) at the bargaining table. Indeed, not all ALMCs are headed down this path. Either due to lack of resources or out of opposition to the underlying principles they have chosen a narrow path of improving dialogue between labor and management on relatively noncontroversial issues. It is important to note that either path reveals the centrality of collective bargaining roles in shaping the evolution of an ALMC. And, for those ALMCs involved in the more far-reaching changes, it is clear that the cooperation enables them to do more than meet traditional industrial relations goals. It also enables them to respond to larger pressures on the industrial relations system.

ALMCs' EMERGING NEEDS

In 1911, Robert Michels first published his classic study, *Political Parties*, in which he observed:

> The party is created as a means to secure an end. Having, however, become an end in itself, endowed with aims and interests of its own, it undergoes detachment . . . from the class which it represents [Michels, 1962: 353].

This iron law of oligarchy, as Michels dubbed it, has endured as an implicit standard in the evaluation of any organization. Viewed from this perspective, we find a constant tension in ALMCs between the need to survive as organizations and the need to accomplish their core mission.

This tension is most clearly manifested in ALMC decisions with respect to funding. At the onset, especially if an ALMC is formed in the face of a crisis, labor and management leaders will play an active role in the organization. Over time, however, the leaders of many ALMCs seek funding in order to hire staffs. This staff, as one report observes, is seen as vital "in view of the fact that labor-management committees are composed of members who hold full-time jobs and who serve on the committee voluntarily" (National Center for Productivity and Quality of Working Life, 1978: 14). However, for most ALMCs, such funding has historically been short-lived or elusive. To understand this critical element of the ALMC story, it will be helpful to review the range of sources to which ALMCs have turned.

In framing the debate over government funding for ALMCs, U.S. Representative Richard Kelly (D-FL) wondered during public hearings, "If this is such a red-hot idea, why do we need . . . the public's money?" He argued that any program that promised to increase productivity and create jobs should have no trouble being supported by the parties (Leone et al., 1982: 227). In fact, testimony at these hearings and analysis in the U.S. Department of Labor's 1982 ALMC study indicates that it is very difficult for traditional adversaries to fund a committee when it has not yet proven itself, and it is even harder to balance the importance of equal contributions with the parties' differing ability to pay (Leone et al., 1982: 128). Put another way, if ALMCs are considered a useful public-policy objective, waiting for self-funding will result in the establishment of fewer ALMCs with comprehensive programs, and, where they are established, it will probably occur after substantial crisis pressures mount.

In seeking outside support, ALMCs have persuaded a variety of federal programs of their merits, including U.S. Department of Commerce's Economic Development Administration (EDA), U.S. Department of Labor's CETA program, and the Federal Mediation and Conciliation Services Grant Program under the National Labor-Management Cooperation Act. In addition, substantial federal funds have been channeled through the Appalachian Regional Commission. However, only the FMCS program is still funding ALMCs and that source of support is limited by its focus on many other forms of cooperation and by its emphasis on short-term seed support. In the future, the recently established division of cooperative labor-management relations at the U.S. Department of Labor is likely to have some additional funds available for established ALMCs, but it is clear that our country's ALMCs will not be sustained by federal support.

At the state and regional level, the Appalachian Regional Commission (with federal funds) and Kentucky state government were, for many years, the only sources of support. Now, ALMCs have successfully lobbied for formal statewide programs of direct support in Pennsylvania and Michigan. With respect to government support, the surest sign of ALMC success and perhaps the greatest hope for the future lies in the growing support these organizations are receiving from city or council coffers and through the sale of services and publications.

Finally, some innovative sources of labor and management support have emerged in a handful of well-established ALMCs. These range from Muskegon's membership structure, based on union membership size and corporate work force, to Cumberland's arrangement with a few local unions and employers to have them contribute one cent per hour (per employee) to the budget of the ALMC. Most organizations do not have such programs, and even these cover only a portion of an ALMC's annual budget, which the U.S. Department of Labor study found to average around $83,000 (Leone et al., 1982: 122).

Although private foundations have provided relatively little support for ALMCs, the above range of sources is still broad. They have, however, generated a variety of program pressures that have forced ALMCs to bend to fit funding sources, or, as Jim Martin, executive director of the Philadelphia ALMC, puts it, to become "market driven" (Martin, 1983).

Still, as ALMC staff have found in Muskegon, Kenosha, Jamestown, and elsewhere, the continued enthusiasm and commitment of labor and management ultimately drives the organization. These ALMCs have undergone the reevaluation needed to modify their programs to achieve the goals both of maintaining funding and sustaining member-support. Sometimes, this has included direct member-contributions, but more often it has involved labor and management leaders creating and maintaining line-item support in city and county budgets. After three to five years of start-up funding, it is appropriate to expect increasing portions of an ALMC's budget to be covered by local "permanent" sources.

Although the evolution leading to such permanent funding is a difficult one that sometimes leads to high ALMC staff turnover, it does provide evidence for a source of hope that Michels (1962) offers in relation to his iron law of oligarchy. He notes that we may never actually discover true democracy, but by continuing the process of searching for it we will, in fact, be achieving it (Michels, 1962: 368). ALMCs may never become pure expressions of labor and management—free of organizational needs—but by continuing to stress this role, they do, in fact, serve this function in society.

FORCES OPPOSED TO THE ALMC

Though much of ALMC activity is driven by an affirmative interest in the accomplishment of various goals and constrained by organizational needs, its path is also affected by a variety of external community forces. Examining these forces and their impact provides an important measure of the extent to which an ALMC realigns power in a community. It also offers insight into (1) why cooperative efforts are not begun in some communities, and (2) why some efforts do not survive. Typically, the external community forces are opposed to the ALMC for economic, strategic, or political reasons.

In one midwestern community, where almost every negotiation for many years involved a strike, one of the most influential local citizens had become quite wealthy through success in various businesses, including the provision of janitorial and other services to companies during strikes. Surprise over his vigorous opposition to the ALMC led to somewhat of a parting of the ways in the management community.

In other cases, ALMCs attract "up-and-coming" leaders from both labor and management. Here, the organization becomes symbolic of internal union or management debates over strategies for the future. Old-line leaders who sustain losses in such a debate sometimes prove a continuing source of opposition to the ALMC.

Although many ALMCs have been able to overcome sources of opposition, others have been less fortunate. In Clinton county, for example, there was strong local government support of the ALMC so long as it was funded with federal EDA and CETA monies. When these funds dried up, however, county commissioners balked at supporting an organization in which, among other things, the executive director earned over twice their salaries. Though donated resources enabled the organization to maintain some presence in the community, it was never again able to offer its previous range of college courses, seminars, conferences, and support to work-site committees.

Eric Trist observes that the task of any ALMC is to demonstrate its distinctive competence and that its power is complementary rather than invasive (Leone et al., 1982: 16). Although this ideal has been achieved in a variety of communities, it has occurred not just as a result of the merits of ALMCs, but by virtue of savvy leadership and realigned economic power.

Shifting power is complex and occurs on many levels. Sometimes, as Bachrach and Baratz (1970: 16) note, the most important exercise of power in a community lies not in making decisions, but in shaping the alternatives. Thus in a number of ALMC communities, the parties, in talking to one another, are surprised to discover the systematic lack of local investment by local banks. In many cases, having this pointed out discreetly by a coalition of labor, management, and political leaders is sufficient for local banks to reappraise their investment policies. In other cases, this remains a source of tension.

Pro-active opposition to an ALMC might be expected from elements within a conservative power-elite in communities with a weak labor movement. In communities with a dominant labor movement, it might be expected that some unions would actively oppose an ALMC as a force that would widen the circle of political influence. In each case, the opposition would be expected to openly refuse to discuss communitywide issues and to exert peer pressure on those who do. In fact, there is little documentation of such behavior. Moreover, the only comprehensive data on failed, joint, communitywide initiatives seem to indicate such pro-active opposition is less significant than other factors.

Five of the eight communities in the U.S. Department of Labor study had joint initiatives at one time or another in the three decades preceding the establishment of the ALMC—an experience that is reflected in many other ALMC communities. The Department of Labor study found that, in most cases, there was little awareness of the earlier initiatives. The demise of these initiatives was attributed to the lack of equal representation from labor and management, the lack of funding and staff, and the lack of extensive programs—especially work-site committees and formal ties to community economic development (Leone et al., 1982: 57-61). Today, even with the availability of basic guidelines for the establishment of an ALMC (covering the previously cited issues), the failure of an initiative to endure or the lack of any initiative is still likely to be explained by factors other than pro-active opposition. Given the dependence of ALMCs on committed leaders, funding, and certain community characteristics, a local stakeholder can often stifle an initiative merely through an ostensibly risk-free path—inaction.

FORCES BEYOND THE COMMUNITY

In discussing a 1940s joint initiative in the British coal mining industry, Eric Trist recently recalled how a postwar consensus to improve production was shattered by a decline in the demand for coal and the sharply differing public policy responses of top labor and management. For the joint initiatives in various mines, Trist (1982) commented, "It was as though the entire ground had shifted under us." ALMCs, too, are subject to and the products of larger societal forces. We have seen, for example, how important funding is to the effectiveness of an ALMC and the extent to which federal funding has been highly variable. We have also seen how the increasing number of ALMCs formed in the 1970s and 1980s is, in part, traceable to shifts in the economy, declining industrial investment, a growing network of third-party neutrals, and an underlying transition in the practice of industrial relations.

At this point, it will be helpful to identify a few other macro forces that influence the establishment and operation of an ALMC. One such factor lies in what Daniel Yankelovich identifies as an undiminished commitment to the work ethic and an increasing demand for a say in workplace matters on the part of the U.S. work force (Yankelovich and Immerwahr, 1984: 61). Whether this is traced to an increasingly educated work force,[10] changing attitudes toward a variety of social institutions as a result of various social movements (Landen, 1982), or other factors, it is clear that the current workplace participation movement is having an impact on ALMCs. In some respects, it adds legitimacy to the work-site cooperation many ALMCs seek to foster. This is especially true where the principal unions in the ALMC are the UAW, CWA, Steelworkers, and others where key leaders have publicly expressed support for participation. On the other hand, where the IAM is a principal union, its opposition to such participation has led some ALMCs to focus on other issues, and it has put some local IAM leaders in the uncomfortable role of departing from the international on this issue. Also, it is important to recognize that union stances are increasingly shaped by corporate strategies that seek a union-free environment. While ALMCs may play a neutral role in relation to labor and management, it should be noted that they cannot be neutral about the importance of a strong industrial relations system in this country.

A final macrofactor to note is the very existence of a growing number of ALMCs. Not only does the existence of prototypes minimalize the risk in establishing a new ALMC, the growing network of ALMCs has pioneered various funding sources.

CONCLUSION

Area labor-management committees are multifaceted, complex organizations. Each is a reflection of the unique characteristics of its community, the forces that led to its creation, the personalities involved, the parties' evolving needs (especially with respect to collective bargaining), its own needs as an organization, the forces opposed to the ALMC, and a variety of larger, societal forces. Only by considering all of these factors is it possible to understand why ALMCs operate as they do.

Such an analysis reveals the important role of a crisis in prompting labor and management leaders to step out of their traditional roles and the extent to which crisis roles can extend into joint organizations that are established. Moreover, it becomes clear that this process is likely to begin *and* develop only where there are strong leaders, third-party facilitators, and relatively equal power relations between labor and management. Within that context, the size of the community, the particular industry mix, and population demographics all serve to constrain or direct the activities of a joint organization. Over time, the emerging needs of the organization (particularly for funding and staff) and opposition to the joint initiative proves a further set of parameters. If successfully addressed, these two parameters expand to allow for a potentially vast impact. Finally, all of the activities of such an ALMC are intimately intertwined with the participants' primary industrial relations roles and deeply influenced by larger social forces. Indeed, the parties' very need to form community-level coalitions can best be understood in light of one central, overlapping larger force—the current transition in the U.S. industrial relations system.

Thus, although each ALMC plays a unique role in its community, all share a common commitment to providing a forum for communitywide dialogue between labor and management. Indeed, Robert Keidel (1978) has documented that an ALMC provides for a diffusion of innovations (such as gainsharing, employee participation, or skills training) that spans organizations rather than merely spreading from one part of an organization to another.

This networking role is at the heart of the contributions ALMCs are making to the practice of industrial relations and local economic revitalization. Yet ALMCs are rarely the subject of academic inquiry or public debate in either field. Certainly there are some scholars and policymakers in the field of economic revitalization who genuinely do not understand just how significant and difficult it is for labor and management to work together. Similarly, there are those in industrial relations who just do not see activity outside of union organizing, collective bargaining, and grievance handling as important. Then there are some in both fields who are opposed to such cooperation for philosophical or doctrinaire reasons. But for the rest, the grass-roots movement is deserving of closer analysis.

Twenty years ago, Chamberlain and Kuhn (1965: 433) observed that the real barrier to cooperation is "paradoxically, a fear of cooperation

itself." Even today, labor and management are not resolving this
paradox out of a sudden zealous commitment to cooperation, but out of
a pragmatic recognition of the importance of problem-solving committees as an adjunct to the bargaining process. Embedded in this
pragmatism, however, is a strong affirmative commitment to improving
the effectiveness and minimizing the externalities of collective bargaining, as well as a growing recognition of strength through coalition. Thus
in the very act of stepping out of traditional industrial relations roles to
contribute jointly to economic revitalization, labor and management
are enhancing their power and developing a much needed mechanism
for the self-monitoring of the industrial relations system. Although the
organizations they have created for this role are still fragile, their
endurance and diffusion portends a significant addition to our nation's
political economy.

NOTES

1. The various vignettes and facts about ALMCs in this article are drawn from my
work as a consultant to ALMCs in Jamestown, New York, and Lock Haven,
Pennsylvania; research with Richard Leone on a nationwide study of ALMCs conducted
for the U.S. Department of Labor; experience in organizing new ALMCs while working at
the Michigan Quality of Worklife Council; associations with ALMC leaders via the
National Association of Area Labor-Management Committees; and the current (though
modest) literature in this field.

2. Such large social problems are sometimes referred to as "meta-problems" or by the
French, "problemtique." Russel Akoff's appelation is, perhaps, more illuminating. He
refers to such a situation as "a mess" and he sees organizations and our society as
increasingly confronting a host of "messes," all with complex and intertwined sets of
causes and elements. In response, he, Trist, and others advocate a "systems" approach to
social problems.

3. Cities or communities with ALMCs will be listed without identifying states; state
references are listed in Table 6.1

4. Although this more formal program might be explained by the existence of
well-publicized ALMC models, such as Jamestown, all other ALMCs established at the
same time or later in the face of a crisis have still begun with a more informal stage.

5. At least this is what I have observed, especially in efforts to organize ALMCs in
over a dozen Michigan communities.

6. For current literature on the subject, see Whyte et al. (1983), Gershenfeld (1982),
and Clark (1982).

7. This historical interpretation represents a departure from most research on labor-
management organizations, which attributes endurance primarily to crisis pressures.
While there is insufficient space here to develop this argument fully, it is pursued in more
depth in Cutcher-Gershenfeld (1985).

8. This is based on statements by many federal mediators, including George C. Happ,
George O'Keefe, Samuel Sackman, and George Vogal, 1982-1983.

9. As Kochan (1980: 415) observes, the system generally lacks any such mechanism.
Also, theorists in other disciplines are increasingly recognizing the importance of such
self-referential behavior (see Hofstadter, 1980; Ospensky, 1950).

10. In 1980, 31.1 percent of persons 25-years-old and over in the the civilian labor force had one or more years of college education. By contrast, in 1959, only 18.6 percent of that group were similarly educated.

REFERENCES

BACHRACH, P. and M. S. BARATZ (1970) Power and Poverty: Theory and Practice. New York: Oxford University Press.
BLONDMAN, M. (1978) "The development of community labor-management committees." Ph.D. dissertation, Cornell University, New York State School of Industrial and Labor Relations.
CHAMBERLAIN, N. W. and J. W. KUHN (1965) Collective Bargaining. New York: McGraw-Hill.
CLARK, S. G. (1982) A Guide to Area Labor-Management Committees. Washington, DC: Appalachian Regional Commission.
CUTCHER-GERSHENFELD, J. (1985) "Reconceiving the web of labor-management relations." Proceedings of the Industrial Relations Research Association, Spring Meeting.
———(1984) "Labor-management cooperation in American communities: what's in it for the unions?" Annals of the American Academy of Political and Social Science 473 (May): 76-87.
GERSHENFELD, J. E. (1982) "Steps in a process: establishing an area labor-management committee." Work Life Review (November).
———and C. G. COSTANZO (1982) A Decade of Change: The Ten-Year Report of the Jamestown Area Labor-Management Committee. Jamestown, NY: Jamestown ALMC.
HOFSTADTER, D. R. (1980) Godel, Escher, Bach: An Eternal Golden Braid. New York: Vintage.
ICHNIOWSKI, C. (1984) "Industrial relations and economic performance: grievances and productivity." Working paper 1567-84, MIT, Sloan School of Management.
KATZ, H. C., T. A. KOCHAN, and K. R. GOBEILLE (1983) "Industrial relations performance, economic performance, and QWL programs: an interplant analysis." Industrial and Labor Relations Review 37 (October).
KEIDEL, R. W. (1978) "Theme appreciation as a construct for organizational change." Management Science (November).
KOCHAN, T. A. (1980) Collective Bargaining: From Theory to Policy to Practice. Homewood, IL: Irwin.
———and L. DYER (1976) "A model of organizational change in the context of union-management relations." Journal of Applied Behavioral Science 12 (November): 59-78.
KOCHAN, T. A. and M. J. PIORE (1984) "Will the new industrial relations last? Implications for the American labor movement." Annals of the American Academy of Political and Social Science 473 (May): 177-189.
KUHN, J. W. (1967) "The grievance process," in J. T. Dunlop and N. W. Chamberlain (eds.) Frontiers of Collective Bargaining. New York: Harper & Row.
LANDEN, D. L. (1982) "Evolution of quality of working life and related trends." General Motors Corporation. (chart)
LEONE, R. D., M. F. ELEEY, D. WATKINS, and J. E. GERSHENFELD (1982) The Operation of Area Labor-Management Committees. Washington, DC: U.S. Department of Labor.
LEWIN, K. (1947) "Frontiers in group dynamics." Human Relations 1: 5-41.

MARTIN, J. (1983) "Creative funding mechanisms." Presented at the annual meeting of the National Association of Area Labor-Management Committees, South Bend, Indiana, November 14.

MICHELS, R. (1962) Political Parties: A Sociological Study of the Oligarchical Tendencies of Modern Democracy. New York: Free Press. (Originally published 1911).

National Center for Productivity and Quality of Working Life (1978) Establishing a Community-Wide Labor-Management Committee. Washington, DC: Government Printing Office.

OSPENSKY, P. D. (1950) The Psychology of Man's Possible Evolution. New York: Vintage.

SIEGEL, I. H. and E. WEINBERG (1982) Labor-Management Cooperation: The American Experience. Kalamazoo, MI: W. E. Upjohn Institute for Employment Research.

TRIST, E. (1982) Videotaped interview with the author. Toronto, June 2.

————(1978) "New directions of hope: recent innovations interconnecting organizational, industrial community, and personal development." John Madge Memorial Lecture, Glasgow, Scotland, November 3.

WHYTE, W. F., T. H. HAMMER, C. MEEK, R. NELSON, and R. N. STERN (1983) Worker Participation and Ownership: Cooperative Strategies for Strengthening Local Economies. Ithaca, NY: ILR.

YANKELOVICH, D. and J. IMMERWAHR (1984) "Putting the work ethic to work." Society (January/February).

7

ACHIEVING LABOR-MANAGEMENT
JOINT ACTION

Warner Woodworth

This chapter reports on a field experiment in which various interest groups in a community attempted to affect problems of regional socioeconomic deterioration. The setting was the small, midwestern community of Muskegon, Michigan, which was plagued by a series of problems. As the Great Lakes shipping industry had declined, the economy began to diminish and relations among sectors of the community rapidly degenerated. In 1971, for instance, 33 percent of the community's total work force went on strike during at least a portion of the year. Management blamed the unions for making impossible demands that were "killing the economy." Labor responded that management had sucked the lifeblood from the area and was now trying to escape. Nine major industries in the area prepared to close by either

AUTHOR'S NOTE: This action research project was supported by the Industrial Expansion Commission (INDEX), of which the author was a consulting participant. Appreciation is expressed to David Cherrington, Paul McKinnon, and J. B. Ritchie for their comments on an earlier draft of the manuscript.

going out of business or moving to the South. The public furor over the two warring factions mounted. Potential new industries, upon learning of the feuds and economic decline, selected other sites for expanding businesses.

The downtown area was boarded up. Urban renewal efforts failed. Plans for a new, central-city mall were abandoned, and the remains of those two aborted efforts made the downtown area look like a fallout region. There was a general exodus from the community—not just from the city, but from the county as well. Statistics indicated that 21 percent of the youth as well as significant proportions of people in other age groups up to age 64 were moving from the area. Signs of decay were everywhere. There were also indications of racial conflict as well as the labor-management conflict. The news media, particularly the local press, were perceived as exacerbating the divisiveness. They were seen as not simply reporting, but as blaming, escalating, and irresponsibly accusing. Depression, both economic and psychological, took its toll, and in frustration people began to express deep-seated fears about their lives, property, jobs, and future.

At this point, a new, fairly neutral organization, funded by banks, utilities, government, and industry, was established to deal with economic expansion. This new entity attempted to begin altering the pattern of decline and despair. A board, consisting of key business and political figures, was formed and it hired an executive director who determined that before they could attract new industry to the area, a number of the existing problems would have to be eradicated. It was reasoned that only a healthy community climate would appeal to new industry. A necessary precedent to obtaining new business would be to retain existing firms. The crux of this challenge was seen to be the resolution of disputes between management and labor. An effort was made to generate "creative, productive dialogue with leaders of the two opposing forces." However, the director was seen by labor as a tool of management, so a decision was made to solicit outside consultants to assist in the complexities and politics of this adversary condition.

The upshot was the formation of a team of consulting specialists. Table 7.1 conveys the flow of events that followed over the next several years. The team that was formed reflected a clear concern for a representational mix of consultants. Two of the consultants were individuals of considerable experience who had worked with different companies nationally, had spent their early years organizing workers in the automotive industry, and who had been active with different groups involved in social change including churches, schools, and communities.

TABLE 7.1
Sequence of Interventions in Development of Joint Labor-Management Efforts

Dates	Event	Stage of Planned Change
Past decades	Economic depression, union and management conflicts, community fragmentation.	System tension, pain.
March 1972	Industrial Expansion Board determines need to improve business climate.	Impetus to act.
April	Team of consultants formed; three-phase plan designed and approved.	Obtain outside help; initial change design.
May	Steering committee of labor and management formed to direct consultants and participants; begin press releases—Project Priority.	Build commitment; secure leadership.
June-July	Phase 1—Interviews with forty top union and industry managers—period of self-examination and exploration of issues as seen by the groups separately. Preparation of reports.	Data collection and analysis.
August	Phase 2—Separate meetings with labor group and management group to assess report of their own interviews, clarify focus, get consensus on labor's position, management's perspective.	Scrutinize issues/strengthen in-group solidarity.
September	Phase 3—Bring labor and management together for exposure to the others' critical concerns. Agreement on joint priorities and establishing of mixed task forces to address those issues.	Identification of superordinate concerns; begin problem-solving strategies.
September, October	Interim work of task forces. Second joint meeting of labor and management. Agreement to co-fund change activity by unions and companies in multipronged thrusts as Phase 4.	Movement from analysis to action.
Fall 1972 to 1982	Coordinating Committee	

Community Forums Committee to Improve Media Overall Economic Development Programs Committee on Governmental Consolidation Contractors/Trades Council Manufacturing In-Plant Forums

A third consultant was a researcher with sophisticated background in data collection and analysis. The other two consulting team members were from academia, with recent Ph.D. work in organizational behavior. The mix included black and white, researchers and activists, conservatives, liberals and radicals, young and old. The team split along the lines of personal values, with two working as advocates of the industrial owners and managers. Two other consultants aligned themselves with and articulated the interests of the labor leaders in the community. The fifth member of the team attempted to coordinate the overall process of diagnosis and intervention.

A proposal from the consulting team entitled "Project Priority" was presented to the industrial expansion board in early 1972. It called for a program that would "identify the causes for the lack of industrial development and the depressed condition of labor and will unite the self-interest groups in the community in moving on a program of positive action." The comprehensive plan consisted of three phases. The first phase was to organize and conduct interviews with representatives of major self-interest groups in the community to identify existing problems. Five sectors were originally designed to be a part of that interview process: labor, management, government, minorities, and an independent category that consisted of schools, churches, professional groups, and so on. Ultimately, only the groups of labor and management were selected. The total group of five was perceived as too big a package to handle—too expensive, too time consuming, too complex. In phase two each group would evaluate the findings from its own series of interviews, analyze critical issues needing solutions, and identify issues it might be willing to work with others to solve. The third phase was the evolution of joint consultation, involving all parties to review data from the other sectors and plan positive steps to improve industrial development and labor opportunity in the area.

The plan was eventually approved and funded by the industrial board. Project Priority became the mechanism for affecting the relationship of labor and management in the community.

The next step was the creation of a steering committee to manage the project—a move designed to ensure responsibility and ownership of the plan and consulting efforts by the client groups, emphasizing involvement of the people themselves. To make up the committee, the change agent team held separate meetings with the top twenty industrialists and top twenty union leaders in the area to explain and test for their interest and support of such a program. Thus both parties were invited to begin a new process simultaneously. This synchronization of participation seemed to be crucial to a sense of joint determination by the two groups.

The industrialists nominated three of the heads of major firms in the area. Labor representatives, after considerable debate, selected a key officer from each of the three union councils: AFL-CIO, Building Trades, and UAW. These six individuals then formed an important structure in the overall change effort, that of a citizen steering committee.

It should be noted that accomplishing these tasks was a complex and arduous task. The realities of designing, formulating, and obtaining commitment to a problem-solving process were difficult. The efforts of consultants were met with doubt, suspicion, and hope. They were not just selling a solution, but an organizational process for analysis and confrontation of issues. Yet, after more than twenty years of decline, of a "growing sense of urgency" as one put it, of "a community falling apart at the seams," as the press put it, there was a willingness to experiment. Clearly the crisis of the economy and the pain of both management and labor were crucial to the tentative agreement to work on the problems they faced. Polarization of perspectives was deep, not only between the two major forces in the economic community, but within them as well—divided labor, isolated and competitive business leaders. The resulting utter lack of accord had produced a systemic paralysis. Project Priority offered an alternative to a community that was dying, not with a bang, but with a whimper.

STAGES OF DEVELOPMENT

The data collection procedure of phase one consisted of consultants' in-depth interviews with some forty key figures of the region's economy. Confidential probing occurred relative to the industrial climate, its history, and current perceptions. Emphasis was on careful exploration of the issues and on searching for underlying causes of conflict and not just their symptoms. Issues were redefined in light of additional information. Views of not only the problems but strengths and potential resources that could be mobilized to address those problems were also researched. In sum, the data collection effort became a tough, frank, painful self-examination, monitored and moderated by the consulting team. The data were synthesized, analyzed, and summarized in a report to each group. In other words, labor received its report on how labor saw the picture, and management received its own as well.

Subsequent meetings were then held in phase two to review the analysis, clarify and attempt to resolve intragroup differences, sharpen the issues, and select top priorities on which to work with the other party. Essentially, this activity was that of creating in-group solidarity, coalescing around common concerns to reach the roots of the dilemmas facing each group as a group, whether it was labor or management. On both sides, it was clear they had some of their "own housecleaning to do" before they could have a productive dialogue with or work with the other party.

Two other outcomes of this important technique may be useful to mention. One was that skilled and insight-producing interviewers developed credibility and legitimacy in the research process. The consultants gained respect, as well as information, throughout the data collection. And this credibility as an outcome was crucial to consulting effectiveness in the later stages of managing intergroup conflict productively. The other important factor flowing from the separate meetings was that each group alone, without the presence and resulting threat of the other, could blow off steam, shout expletives, and sort out peripheral issues that might sideline the debate over real problems and substantive issues later. Thus with time, in their own meetings, each group, management and labor, was able to work to target basic difficulties and build consensus on core concerns that would be more likely to produce a willingness from the other part to listen, take the other seriously, and begin truly to understand the other. By this process of focusing they were able to develop a case for each issue that would invite constructive response.

Table 7.2 lists the dominant concerns of each group. Overall, it appears that a major difference between the tone of the two reports was that the thrust of management tended to be on organizational needs, company survival, and profitability. Labor, on the other hand, seemed to emphasize more the human needs, the concerns of the individual in the street.

The phase three intervention was that of a joint consultation that would bring management and labor together for the first time, totaling forty economic leaders from different plants and institutions. The consulting staff had met with spokespeople designated by each group during the previous week to ensure success at this initial meeting. Emphasis was given to presenting one's group issues as accurately as possible, and on articulating their position in ways that would least provoke reaction or prevent further discussion. A one-day meeting was held with the forty participants sitting in groups of five at tables scattered throughout the room. At the outset, the senior consultant and coordinator of the event stressed to all that "our objective is not to jump to conclusions. The main goal today is to reach some joint under-

TABLE 7.2
Differing Perceptions of Labor and Management

Unions	*Management*
Union View of Management	*Management View of Unions*
(1) We are in a crunch. Things will come to a boil unless changes occur.	(1) Workers are just out to screw the company.
(2) Too much absentee ownership— no sense of social responsibility.	(2) We are forced to pay high labor costs and fringe benefits—way above normal.
(3) Local firms being bought up by conglomerates.	(3) Labor is too hostile to management, too militant over the years.
(4) Management intimidates labor leaders. We never get access to top management.	(4) Workers resist changes that might increase productivity and fight to maintain crippling work practices.
(5) All is relegated to the IR directors.	(5) The area labor force is overprone to use the strike.
(6) Contractors are losing too many bids to outsiders, result is little local work.	(6) Labor relations are too unpredictable here, so we are moving our company elsewhere.
(7) Management won't bury the hatchet.	(7) The international union interferes too much in our plans.
(8) Labor is intentionally excluded from management's plans.	(8) We are forced to inherit the UAW pattern set in Detroit.
(9) Both groups need to communicate more—not just at contract time.	
(10) We can't believe management is sincere.	
(11) Management is too greedy.	
Union View of Itself	*Management View of Itself*
(1) Labor needs to get better organized—need for more team spirit.	(1) We need to win back management rights bargained away in past years.
(2) Unions need foresight not to overkill at the bargaining table or to drive companies away.	(2) There ought to be improved forms of communication with our employees—in plant newsletters, small group meetings, regular discussions with union officials between negotiations, etc.
(3) There is conflict among our people, black/white/male/female, outside/local unions, etc.	(3) There are too many inducements to relocate in the South: 25 percent reduced labor costs, tax incentives, etc., and there is nothing to attract us to stay here.
(4) Declining in membership—need to change our image.	
(5) Rank and file are not involved any more.	(4) We need higher productivity in order to pay high wages.
(6) We can't fraternize with management or we'll lose members' support.	(5) Our competitors are killing us, especially in low fringe benefits. We just can't survive. Our competitive edge has eroded and we have got to act.
(7) The tight squeeze causes jealousy and infighting among locals.	
(8) Management and labor are on different sides of the fence. We must never forget that.	(6) A further challenge many companies face is that of plant obsolescence. Our buildings and equipment are too old.

(continued)

TABLE 7.2 Continued

Unions	Management
Union Ideas About Other Issues	*Management Ideas About Other Issues*
(1) The community is sick and dying.	((1) The media don't present economic
(2) People are too isolated. Everyone is concerned only with self, not the needs of others.	issues to the public. Lack of informed community creates misunderstandings.
(3) The news media treats labor in a derogatory fashion and makes management-labor disputes look worse than they really are. It's too divisive.	(2) The community lacks any unifying thrust to pool its total resources.
(4) There's no future for our youth. They turn 18 and leave home, or stay and rot.	(3) Our image in surrounding areas is totally negative and this breeds pessimism back home.
(5) City fathers are looking backward, dragging their feet. They will do nothing until the city erupts.	(4) There is lack of coordination between industry and government. Government does not understand management, nor does it support new industry. The result is decay.
(6) Too much fragmentation and wasting of resources by government units in the area.	(5) There is a lack of community esprit de corps.
(7) Racial oppression is high. Minorities are the last hired and first fired.	
(8) Blacks are being locked into the central city, poor housing, 35 percent out of work, etc.	
(9) People are afraid. They are arming themselves for protection, voting for George Wallace en masse, and so on.	
(10) There is a lot of untapped economic potential here: waterfront, resort development, downtown renewal, nuclear power plant, etc.	

standing of the problems we face. If we can do that, then we can select joint groups for moving towards solutions."

To reduce the likelihood that open conflict might disrupt the meeting at the outset, those presenting their respective sides were invited by the consulting staff to first discuss those issues on which labor and management shared some agreement. Thus initial presentations were rather safe, dealing with criticism of local news media and the need for better communication, problems of governmental and community fragmentation, increased participation in civic affairs, and so on. They then approached the "gut issues": management's loss of its competitive position and labor's concern with destructive union-management relations. The ensuing discussion pushed in both directions. There were concessions on both sides, as well as considerable stereotyping and nonlistening. Yet there emerged a general feeling that changes must occur and that the two groups needed each other. It was also felt that there existed enough common concerns to warrant continuing the dialogue. After lunch, a number of areas were identified as critical and potentially workable by some form of joint action. The priorities selected for intensive focus were as follows:

(1) improved communication between top management and union leaders—better working relations

(2) greater efforts by the news media to communicate accurately and fully the economic realities of the region

(3) reduction of hostility and divisiveness among sectors of the community—labor, management, government, and so on

(4) joint support for the execution of an overall economic development study that proposed certain growth and industrial activity in the area

Upon agreeing to these priorities, the forty participants split into subgroups, with consultants as facilitators, to organize themselves as task forces that would work on each of the four concerns. There was a mix of representatives from both labor and management on each team. In subsequent weeks the groups met, collected data, debated alternatives, and prepared reports for a second joint consultation. At the second meeting recommendations were proposed by each team. Some suggested the expenditure of resources for further study. Others proposed plans for short- and/or long-range actions. Each report was discussed by all forty participants, resulting in both affirmative feedback and criticism. In some cases, the task force was urged toward a more aggressive stance. In others, the demand for better information was made.

Proposals that were deemed worthy of further pursuit included initial plans to select specific economic development projects from the overall plan that management and labor could jointly influence most. For instance, they decided to lobby together for the awarding of major new construction projects to local firms, which would secure profits and jobs within the area. Another set of proposals had to do with establishing labor/management forums in various companies to discuss aspects of the working climate that were outside the traditional bargaining process. Another proposal was that communitywide educational events be sponsored by Project Priority, with dinners held so labor and management could socialize and outside experts could come to speak on topics that would increase the understanding of the public, particularly with regard to economic issues. The media group advocated meetings with local press leaders to relay criticisms of news coverage, invite a closer relationship with economic forces in the community, and design informational columns in the paper and news releases sponsored by Project Priority. Contractors and trade leaders announced plans to recruit additional participants and begin dialogue on issues specific to the construction industry of the region. The group dealing with community consolidation had been unable to devise any specific action plans at this point and it became deferred to Project Priority as something they could do little about at that time. Most members of the group agreed with this, as consolidation was really more a governmental issue. Also there was some concern, on the part of consultants especially, that too many issues could dilute the group's energy and it would lose its focus.

The final discussion at this second joint meeting had to do with the hard realities of financial backing for future projects. Project Priority had been funded by the industrial expansion board through the first three phases of (1) interviews, (2) separate meetings and analysis, and (3) joint consultations for the sharing of information and testing for collaborative action. The implementation of these plans and execution of task-force recommendations would require further dollar resources. The board made a quick, rough calculation and concluded that implementing the actions proposed would cost an additional $45,000. Although the board agreed to provide some money to keep the momentum going, it was clear they did not have the resources to fund the total project much further.

Equally important was the feeling of the consultants that participating institutions, both labor and management, needed to pay their own way. The notion was that not only would such a strategy be more feasible in budgetary terms, but psychologically it would create increased feelings of ownership and responsibility in making things work. However sound such reasoning might be, the reality of financial commitment was quite different. Many of the forty participants suddenly got cold feet. There was considerable dissent and the general feeling that they needed to return to their organizations to think this idea through, check with colleagues, and return at a later date to resolve the thorny problem of financial support.

The rather surprising result was that after considerable deliberation between management and their boards of directors or top company officials, and unions with their local leadership council votes, both sides agreed to share jointly the costs of continued consulting services. There was a mixed feeling that considerable progress had been made and/or that the program held enough potential to warrant a more direct financial commitment to the effort. It should be noted that several firms dropped out at this stage, people who did not see a payoff coming from further investment on their part. Also important was the fact that larger firms were assessed a greater dollar participation than small companies according to a Project Priority formula. The inclusion of union funds to such a program was a unique and significant factor in the history of this project in the community, and of change projects nationally.

Thus began a new phase four. At this point, a coordinating committee was established to expand on the earlier steering committee membership. It was designed to provide leadership to the newly developed task forces, to function as a center of communications, to receive feedback on the diverse thrust of Project Priority activities, and to monitor progress over the next months. (See Table 7.1 for information helpful in understanding the structure of events in process at this time, along with the formulation and focus of various task forces.)

With staff assistance and the catalytic efforts of the consulting team, task-force action plans were implemented. As participants learned to work together effectively, as trust was established, and as initial successes were achieved, the consulting staff gradually withdrew from the process. Some task forces worked themselves out of jobs within a few months, either through lack of a substantive set of proposals or by meeting the original goals. Others continued for some time, and their efforts eventually became institutionalized in a more or less permanent form in the community.

PROJECT OUTCOMES

Now, a decade since the project began, an assessment can appropriately be made of the fruits of this collaborative effort by local management and labor. In weighing the outcomes, the data tend to show some important trends since the inception of Project Priority in 1972. The effort existed solely on local funds for companies and unions from 1972 to 1975. At that point the labor/management board applied for an Economic Development Administration (EDA) grant, which they received. The $86,000 grant was spread over three years, allowing the organization to hire a coordinator and secretary as paid staff. Additional funds were obtained from the city and county, and the steering committee became a board, expanded from six to eight members, four from labor and four from management. By 1977 Project Priority had evolved into the Muskegon Area Labor-Management Committee (MALMC); it has blossomed into a classic model for joint problem solving.

Today 36 organizations have some type of participation in this outgoing effort to revitalize communitywide social and economic

TABLE 7.3
Results of Project Priority/MALMC:
A Ten-Year Perspective

Community Forums
- Designed a series of monthly programs to which all sectors of the city were invited.
- Dinner, friendship, and instruction by expert resources on such issues as the economy, production, and racism.
- Committee arranged programs, recruited speakers, raised funds.
- Successful educational forums continued on annual basis through 1982.

Committee to Improve the Media
- Conducted various meetings with three major newspapers, TV, and radio to feed to them joint concerns of unions and companies.
- Specifically advocated less inflammatory editorials, elimination of negative racial references.
- Formulated news releases shared by management and labor in an effort to ensure more balanced coverage of economic issues.

Committee on Governmental Consolidation
- Report was begun on problems of government overlap and assessment of the public's attitude toward metropolitan consolidation.
- After several months it was concluded this task was essentially political rather than economic in nature, perhaps beyond original charter of Project Priority.
- Effort was abandoned.

Overall Economic Development Programs
- Generation of public support for comprehensive economic growth projects.
- Particularly instrumental in building solid waste disposal facilities, attracting three new chemical businesses, completion of new downtown mall as a source of community pride, expansion and success of new industrial park to relocate older local firms within the area and successfully attract new industry.
- Expansion of older firms (creating new jobs) and investing in new, updated equipment and technology.

Contractors/Trades Council
- Team formation to improve competitive position on new bids, reducing bids to nonunion or outside union contractors.
- Dialogue and subsequent reduction of jurisdictional disputes between the trades.
- Redesign apprenticeship programs with emphasis on recruiting and retaining minorities.
- Prejob conferences before construction began on new projects, among various trades, contractors, and superintendents, clarifying problems at outset.
- Monitoring teams, composed of both labor and management, to inspect sites and ensure compliance with OSHA regulations, quality inspection, minority hiring, and wage standards.
- Ongoing dialogue with occasional guest resources to explain OSHA, or architects and engineers to discuss bidding process.

Manufacturing Plant Forums
- Creation of in-plant committees of 5-10 representatives from top management and key union leaders to meet 1-2 times per month.
- Intent not to replace or subvert regular bargaining process in any way—function entirely supplemental on nonnegotiable issues.
- Each plant (12 in all have participated) established its own set of procedures and issues with the help of the consultant monitoring, troubleshooting, etc.
- Sample outcomes include companies that opened their books to labor, letting the union see and understand the decline of the profit picture; unions were included

TABLE 7.3 Continued

in management planning activities with regard to expanding facilities; improved
conditions with respect to worker health and safety; improved job assignments.
• Improved contract negotiations in 1973, 1976, and subsequently: Both sides
attributed the smooth and improved outcomes of bargaining to Project Priority.
Forum structure and dialogue is still going on ten years later.

challenges. Table 7.3 outlines the results of this intriguing experiment in
labor/management reform.

The thrust of the projects is particularly noteworthy in several areas,
each of which is discussed below.

Education. The committee sponsors an annual tribute dinner to
which labor, business, and government leaders are invited to hear about
major issues facing the region, the economic picture, and other topics of
interest. It averages an attendance of 300 to 400 participants. Also,
seminars are sponsored jointly with the local community college on such
topics as absenteeism, substance abuse, and quality of working life.

Communication. Informal round tables are held on a regular basis
wherein various managers and community leaders can gather for
dialogue, to appraise local developments and, basically, achieve an
ongoing relationship of openness. The committee sponsors an annual
conference outside the area, where companies and unions have sent up
to sixty participants for a several-day seminar on issues of common
concern, featuring academic and/or other resources in selected areas of
expertise.

In-Plant Forums. Between eight and ten companies and local unions
(mostly UAW affiliates) have established joint committees in their
plants to anticipate and prevent on-site problems. Beyond efforts to
build trust and improve relationships between the two sides, the forums
have dealt with such problems as scrap reduction, retooling, absenteeism, safety concerns, a new incentive system for workers, and so on. In
one forum, for instance, with input from the union, the company made
design changes in their products, which allowed it to obtain a contract
that had previously been bid on unsuccessfully.

Area Economic Regeneration. Project Priority has worked to build
public support for various economic improvements in the region. These
have included the successful conclusion of the construction of facilities
for the disposal of solid waste, the building of a new downtown
shopping mall, and the expansion of a new industrial park. Improved
relationships between management and labor have begun to alter the
negative image of this region as a "war zone," resulting in decisions by
three chemical companies to relocate in the area. Several old plants have
invested in and installed new, modern equipment, with one automotive

parts plant embarking on a $7 million expansion project resulting in new
jobs for people living in the region. In contrast to double-digit
unemployment of the war years (1971-1972), by 1978 the jobless figure
had dropped to 5.8 percent.

Conflict Reduction. Specific improvements, such as the expanding
plant discussed above, are an outgrowth of a transformation in the
industrial climate of the region. In that case, the plant had a fifty-year
history of labor conflict that included a fourteen-week strike over
disputed contract negotiations in 1972. The contract settlement of 1981
was effected very smoothly, due to the in-plant forum structure for
solving problems. Hence the company determined it would be to its
own, as well as the employees' advantage, to expand its facilities and
create new jobs. In contrast to the one-third of the community's work
force that went on strike in 1971, by 1982 there were no major walkouts
as contracts were settled peacefully. In various plants, work rules have
become more flexible, leading to improved productivity, better product
quality, and lower operating costs.

SIGNIFICANT ASPECTS

To be sure, this change strategy performed no miraculous transformation of all of the region's problems. The 1981-1982 recession hit the
Great Lakes area very hard, but Muskegon weathered the storm better
than Michigan as a whole. A concluding commentary on this rather
intense, prolonged, and complex consulting effort ought to address such
questions as what was learned, what were the significant interventions,
processes, and methods that contributed to the successful outcomes?
The case is unique in its coverage of more than ten years, which allows us
to assess its impact. In retrospect, what aspects of the project seem to
highlight its development? The sections that follow point out some of
the important lessons learned that might suggest strategic implications
for future consulting work with unions and management.

CLIENT DIRECTEDNESS

One central feature of Project Priority was the importance and role of
the steering committee established at the outset. It was a mechanism that
facilitated many tasks the consultants could not accomplish alone, such
as ensuring support from key figures on both sides, serving as a model of

collaboration to the masses of their respective constituents, and initiating a sense of ownership of the project and commitment to it. These committee members were critical to the consulting interventions as a reliable source of input where ideas could be tested. They forced a consulting effort that made the staff more responsible to their clients. The consulting team was pushed by them in an orientation toward people, not toward research or theory per se. They helped formulate effective strategies and kept the change agents honest and politically "on target."

A clear conclusion is that such sponsors should be important personalities in the community, that they should wield influence as decision makers and have solid credibility, that they should represent organizations and not just themselves, that they should be aware of problems, and that they be committed to change. In the case of Project Priority, most of the above criteria were met by Steering Committee members. Additionally, the right chemistry of personal charisma and political savvy occurred as unplanned, but vital, fallouts in the selection process. All in all, the Steering Committee deserves much credit for the eventual success of the project.

EQUAL RESPONSIBILITY

A truly unique dimension of the entire process was the issue of equity and joint responsibility for the program by management and labor. Historically, most community efforts have been initiated and funded solely by the companies involved. It was deemed by the consulting staff that to ensure their not being stereotyped or entrapped into management loyalties, union ownership of the activities had to equal that of management—not only in terms of representation but through the committing force of hard, cold cash. Equal pay for equal service does a lot to remove an elitist stigma to intervention efforts and continuously sensitizes the consultant to his or her professional obligations.

PUBLIC OPENNESS

Another vitally important characteristic of Project Priority was that of "going public" from the outset. In many settings the dominant view is to issue no press releases until "significant news" about substantive outcomes is achieved. Often, when the action plans are initiated a blocking and smothering effort by the power structure occurs and there is a simultaneous cover-up about the investment of time, energy, and financial resources with no public accountability—and no public outcry at the lack of results.

In the case of the experiment of Project Priority, news releases were issued often so the community would be informed of the steps, not just the outcome. Such a procedure was useful for the process too: It kept things aboveboard and honest, it encouraged a common vision or collective image of labor and management before the public, and the increasing series of articles helped maintain momentum.

RESEARCH VERSUS ACTION

Related to momentum, a critical stage at which things might have really become bogged down was after the forming of task forces and their early efforts at analyzing and planning. One observes in many efforts that research-oriented consultants get themselves and their clients stuck in the paralysis of analysis. In this project the concern was that there not be a major regression from action proposals back to lengthy study. From the outset of Project Priority, the change agents billed themselves as committed to action, not just to an academic motive of collecting interesting data. Both labor and management leaders tend to be action oriented and, although the consultant can certainly choose to sit and analyze data while hoping his clients change their minds and come to appreciate the research approach, the prevailing sentiment is to yield results. Many projects are blocked, constipated precisely at this critical juncture.

ANALYSIS

On the other hand, and this is part of the thesis/antithesis process of intervention, it seemed the consultants could not overemphasize the need for time, careful diagnosis, and thoughtful reflection on the issues. Defining and redefining before jumping to solutions was underscored from the outset. The change team felt strongly that the problems in the community were by no means simple. Band-aids would not work. Both groups needed to move beyond hit-and-run disputes of the past to a more preventive stance, beyond blaming to mutual problem solving. Anticipatory problem resolutions rather than post facto crises of the moment demand an understanding of self and others before action can be taken. The results, over time, especially in the case of in-plant forums, contributed greatly to an improved quality of decisions and answers.

TENSION AND PAIN

The experience reaffirmed the Lewinian notion that before change can occur a system has to unfreeze. Unions and management would probably not have collaborated on Project Priority had it not been for the deepening economic crisis in the area. From the beginning, the proposal looked risky to many. Even among its original sponsors, the industrial expansion board, there were feelings of high anxiety. Said one: "Can we take the chance? It might explode in our faces!" Yet the intense pain of present conditions and the resulting high level of frustration prevailed as another spoke: "Can we afford not to try?" Thus as the two parties began to work with consultants, there emerged a clear recognition and acceptance of a superordinate goal (Sherif and Sherif, 1958)—the survival of both groups. Each began to perceive it could gain more by working together toward a win/win outcome than the history of lose/lose in which each was cut off and isolated in the hostile environment of the past. While Kochan and Dyer (1976: 67) assert that "problem solving or appeals to super-ordinate goals are likely to be ineffective" in labor/management conflicts, one would argue that such a proposition in the abstract is meaningless. Theory cannot ignore political and economic environments in which change is to be operative, and in the case of Project Priority, the appropriateness of a subordinate goal is clear. However, it was not "laid on" by the experts, but surfaced naturally out of the context of economic reality.

INTERORGANIZATIONAL INTERVENTIONS

A last unique and significant point to raise with respect to this particular project has to do with the fact that most reported applications of behavioral science to labor/management relations involve the two parties within the context of a single plant (Blake et al., 1965; Lewicki and Alderfer, 1973). Project Priority occurred within the context and complexity of a community, thus involving a total system (or systems). It included a multifaceted series of interventions with various plant facilities and union locals, with conflicting needs and demands in a heavily political setting. Although it may be argued that considerable success was achieved, the events raise serious concerns about the utility of traditional theory in such an environment.

CONCLUSION

While the outcomes of the change effort suggest a series of positive developments, it should be made explicit that the collaborative effect was not one of love and trust between unions and managers. The process discussed above did not result in the two parties merging into one big family that will live happily ever after. Rather, the process sharpened distinctions between the two groups so that they better understood each other. Unity of two opposing forces was not the intention, for such a blurring usually leads to distrust and cooptation of one party by the other.

To avoid potential exploitation, the consulting team continuously worked to achieve a power parity between management and labor—equal representation, with top leaders of both sides involved, equivalent input to the process, and both factions having access to consulting expertise. The resulting balance created a mutual exchange of benefits, a characteristic central to achieving social justice. Thus the pluralism of the adversarial relationship was maintained, but it was redefined so that both groups could cope more effectively with changing socioeconomic realities. Instead of coalescing on goals, labor and management agreed on a new process—a common problem-solving method that would allow each to better succeed in meeting their own unique objectives. In sum, the outcome was not the smoothing over of antagonism but the creation of a new process for handling conflict in a more constructive way.

A final important implication of the field experiment reported here is that neutrality and behavioral science objectivity are largely mythical in intervention practice. What is needed, therefore, is not the mask of an alleged third party, but rather a structured partisanship that allows for the expression of honest value positions. In the activities reported above, a key to success was the opportunity to move from exploitative games by a consulting team trying to be "all things to all people" toward a genuine advocacy of personal views and feelings. Management and labor were thus able to articulate their values, too, in a legitimate approach, first as self-interest groups and then in joint actions with the other party. Such a model is not only more effective in generating substantive socioeconomic changes, but it allows people to maintain their own integrity in the process.

REFERENCES

BARITZ, L. (1960) The Servants of Power. Middletown, CT: Wesleyan University Press.

BLAKE, R. R., J. S. MOUTON, and R. L. SLOMA (1965) "The union-management intergroup laboratory: strategy for resolving intergroup conflict." Journal of Applied Behavioral Science 1: 15-57.

COSER, L. (1956) The Functions of Social Conflict. New York: Free Press.

KOCHAN, R. J. and L. DYER (1976) "A model of organizational change in the context of union-management relations." Journal of Applied Behavioral Science 12: 59-78.

LEWICKI, R. J. and C. P. ALDERFER (1973) "The tensions between research and intervention in intergroup conflict." Journal of Applied Behavioral Science 9: 424-449.

LIKERT, R. (1967) The Human Organization. New York: McGraw-Hill.

PAYNE, P. (1977) "The consultants who coach the violators." Federationist 84: 22-29.

SCHEIN, E. H. (1965) Organizational Psychology. Englewood Cliffs, NJ: Prentice-Hall.

SHERIF, M. and C. SHERIF (1958) "Superordinate goals in the reduction of intergroup conflict." American Journal of Sociology 43: 349-356.

TAYLOR, F. W. (1967) The Principles of Scientific Management. New York: Norton. (Originally published 1911.)

8

LABOR-MANAGEMENT
COMMITTEE OUTCOMES:
THE JAMESTOWN CASE

Christopher Meek

Perhaps not surprisingly, given that major innovations frequently are born out of crisis, the particular revitalization strategy discussed in this chapter emerged in New York State in the small industrial city of Jamestown, which is located in the southwestern corner of the state. The major force behind this city's success has been a joint effort by local union and management leaders through a community-based committee called the Jamestown Area Labor-Management Committee (JALMC). As successful as JALMC has been, it is important to note that it was by no means the first organization of its type. In general terms, some five area-based labor-management committees (LMCs) preceded JALMC, and thirty or more have come into existence since the beginning of the Jamestown program. On a more specific level, JALMC represents a major departure from the majority of its sister LMCs, and for this reason it has attracted so much attention since its inception in 1972.

Visitors from many countries have found their way to Jamestown from such diverse places as Ghana, France, Scotland, and Norway. Management and union representatives of JALMC and its professional staff have consulted with more than 130 different U.S.-based communities and organizations that have expressed an interest in learning more about area-based labor-management cooperation.

This widespread interest directed toward the Jamestown effort has been the result of some impressive achievements that can be attributed only to an approach quite unique compared to that adopted by other community LMCs. Unlike the majority of its counterpart committees, JALMC has not primarily confined itself to a goal of improving labor relations, but rather has sought to develop a broad-based approach to industrial revitalization informed by behavioral science expertise and solidly grounded in a quality of working-life orientation toward organizational improvement. More specifically, JALMC presents the following distinctive features that set it apart from other similar organizations:

(1) cooperative action by union, management, and local government leaders to save jobs in plant shutdowns and to strengthen the economic base of the community

(2) the creation of in-plant cooperative problem-solving projects in which labor and management jointly define, examine, and make decisions on methods for improving industrial performance and the quality of working life aided by JALMC consultants serving in a third-party facilitating role

(3) a community-centered orientation toward plant-level improvement projects, with the intent of building supportive networks across firms and unions in order to facilitate interorganizational learning and the diffusion of innovation

(4) creative linking of the resources of various local, public, and private sector organizations through joint projects and programs designed to meet special community industrial needs

(5) avoidance of direct involvement in mediation of labor-management disputes

The omission noted in the last point deserves further comment, since the first community based labor-management committees in Toledo, Ohio, and Louisville, Kentucky, were started during the mid-1940s for the explicit purpose of providing a locally based mediation service for resolving union-management conflicts.

In contrast to groups such as the Toledo committee, JALMC's leaders made the conscious choice to abstain religiously from the mediating role. This decision forced the committee's members and staff to find new ways of improving union-management relations long before the emergence of a crisis conflict, and required that they develop novel approaches for reviving the city's ailing manufacturing sector through joint action outside the realm of contractually negotiated agreements. To understand how these new methods were discovered, it is necessary to return to the beginning of organizational efforts in Jamestown, and to follow JALMC's growth and development through what we see as at least four distinct and very important phases of learning.

THE PROBLEM AND EMERGENCE OF THE CONCEPT: 1965-1971

An industrial city of approximately 40,000 inhabitants in a metropolitan area totaling 60,000 citizens, Jamestown has about twice the proportion of manufacturing jobs as the average city in New York State. Jamestown's industry is heavily concentrated in wood furniture and metal fabrication, and the furniture industry, especially, has been losing ground since the Great Depression due to plant closures and the relocation of companies to North Carolina.

The majority of Jamestown's industrial plants are what is known as either "small batch" or custom "job shop" operations, and because of this a large proportion of factory jobs require highly skilled personnel. Because a large portion of the manufacturing jobs available require skilled labor, it is not surprising that Jamestown is largely a union town.

During the 1940s and 1950s Jamestown became renowned for its labor-management conflict and it was an era of strikes that has not been surpassed since. The decade of the 1950s, for example, was marked by 62 different manufacturing strikes and the accumulation of 1986 strike days, compared with the 1960s during which there were 39 work stoppages and 1442 strike days accumulated. This contrasts sharply with the 1970s, when there were only 31 strikes and 749 strike days.

The consequence of labor's militancy and tenacity during the late 1940s and 1950s was the realignment of Jamestown area wages, benefits, and conditions of employment to levels more comparable with those in other industrial centers than they had ever been before (McMahon, 1958: 199). This was a significant accomplishment, given the fact that the wages of Jamestown's factories had been considerably lower than the national industrial averages up until that point (McMahon, 1958: 199). Unfortunately, these gains were not achieved without taking a considerable toll in terms of the community's ability to attract new industrial development and stimulate industrial growth. The news of Jamestown's labor-management battles and supposedly militant unions spread across the country, and manufacturers that knew of Jamestown for the most part ruled out any consideration of building new operations in the area. Thus Jamestown in effect assumed a place on the industrial development "blacklist" because of this notorious reputation for having a "bad labor climate" and being a "strike-happy town."

Growing awareness of Jamestown's problems led in the mid-1960s to the first remedial efforts. A local labor attorney, Ray Anderson, began consulting with the area's federal mediator, Sam Sackman, and with Al Mardon, executive director of the Jamestown Area Manufacturer's Association. All three were familiar with the concept of labor-management cooperation from experiences elsewhere, and they felt strongly that a similar approach could help solve the problems of Jamestown. Furthermore, Anderson and Sackman, in particular, carried their beliefs forward into action by organizing several in-plant LMCs in Jamestown during the late 1960s and early 1970s. They also brought manufacturing and union leaders together several times through jointly sponsored conferences, dinner meetings, and workshops. The intent of these activities was to build sufficient enthusiasm and shared appreciation for the concept of cooperation to enable development of communitywide labor-management collaboration on changing the city's image and tackling its industrial problems. However, after considerable effort, it became clear that actually pulling both parties together with the explicit aim of developing an ongoing areawide LMC required both a mobilizing crisis and a third-party actor capable of both staying involved and spearheading the endeavor on a long-term basis. Both needed preconditions emerged at the outset of the new decade.

ORGANIZING THE RESPONSE TO
THE COMMUNITY CRISIS:
1971-1974

A clear crisis suddenly became evident in 1971 with the announcement of the closing of an old and respected Jamestown manufacturing company, Art Metal. The community's largest employer, Art Metal had been in existence for seventy years, with as many as 1,700 employees on the payroll at times.

Nor was the Art Metal shutdown the end. Imminent closings of six more plants threatened to bring the total immediate loss of employment to as many as 2,800 jobs.

In this crisis, Anderson, Sackman, and Mardon met with Sam
Nalbone, city ombudsman and formerly a business agent for the
International Association of Machinists and Aerospace Workers (IA-
MAW). Nalbone brought them together with Stanley Lundine, who had
been elected mayor in 1969.

Lundine was a clear choice for assuming the leadership in the creation
of community-based LMC. In his first campaign he had taken a strong
stand for government activism in solving Jamestown's economic
problems, and also had solid political backing from both labor and
management. Furthermore, he had already discovered the futility of
traditional redevelopment strategies and was searching for a new
approach. As he himself said several years later:

> I had thought, I guess with some degree of naïvete, that we could just go
> out and lure industry into this area, and that all we needed was an
> industrial development program. I went all over seeking new industry for
> the area and it seemed that everywhere I went they said we had a "bad
> labor climate." . . . So I decided that redevelopment in the conventional
> sense just wasn't enough for us. We had to plunge into the economic area
> having decided that it's pretty obvious that Jamestown is a manufacturing
> town. . . . Having analyzed the manufacturing it was very obvious we had
> a highly unionized town. So I felt you had to get the top executives of the
> manufacturing companies and the union leaders to talk about their
> problems. No government program was essentially going to solve it.

Assisted by Anderson, Mardon, and Nalbone, Mayor Lundine met
first with key leaders of labor and management in separate caucuses and
then brought the two groups together. The first joint meeting (February
1972) began in stormy fashion with labor accusing management of
keeping business out of town in order to keep wages low, and
management accusing labor leaders of creating the bad labor image.
After ventilation of these charges, Lundine got the parties talking about
how to improve the situation and they arrived at the approach described
by Lundine below:

> We agreed temporarily that there would be a committee, the labor-
> management committee; we would not form it legally. . . . It would be the
> type of thing where anybody could take a walk anytime they wanted to.
> We agreed that at the next meeting we would choose co-chairmen.

After considerable informal consultation with both management and
trade union leaders, the mayor called the group together for a second
joint meeting in March to choose cochairpersons: for management,
Allen Yahn, plant manager at the local American Sterilizer Company
plant; for the trade unions, Joe Mason, business agent for IAM/District

65. In addition to choosing their leadership, the group also agreed on the following four objectives as the major concerns toward which JALMC would direct its efforts:

(1) improvement of area labor relations
(2) manpower training and development
(3) assistance to local industrial development activity
(4) improvement of industrial productivity through union-management effort.

Commitment of the unions to the productivity objective required extensive consultation outside of joint meetings. Labor leaders were naturally suspicious of a goal that might mean speed-up and loss of jobs. A key premeeting discussion between Lundine and Joe Mason ended with this statement by the labor leader, as recalled by Lundine:

> Okay, productivity becomes a major objective, but, one, we want an understanding that it's not going to be used to cut out workers. If we're going to improve productivity in a plant, then management is going to do everything possible not to lay off workers. Two, we want a general sort of understanding, without any sort of negotiating up front, that gains will be shared equally by labor and management, and three, we want to define it broadly, we don't want speed-up. Now those weren't all Joe's ideas: those were ideas that kind of developed in those side meetings. On that basis Joe was willing to accept the chairmanship.

The March meeting also led to an agreement on ground rules: Membership on the committee was to be entirely voluntary, and there would be a balance between labor and management, both in terms of leadership and of membership. JALMC would avoid any direct involvement in collective bargaining beyond developing special training courses and workshops.

The mayor and ombudsman were designated ex-officio members of the committee, which underlined their commitment to decision making by labor and management, but placed them in strategic positions to work behind the scenes to keep the parties moving together.

JALMC's first task was to save 300 jobs in the impending shutdown of the Chautauqua Hardware Company. This old, locally owned firm had been mismanaged to the point that productivity per worker measured 60 percent of the industry average. The mayor and local union leaders persuaded the bankruptcy court to allow the plant to keep operating. With advice and assistance from other committee members, the mayor went about finding potential investors and managers. He also brought the potential new owner-managers together with local union leaders to reach an understanding on future cooperative efforts.

In bargaining the first contract in the fall of 1972, management negotiators announced that their unconditional offer had to be limited to a 4 percent cost-of-living adjustment but promised productivity bonuses. The parties then negotiated the formula for such bonuses, thus signing Jamestown's first productivity agreement.

No explicit productivity agreements were reached in other threatened shutdowns, but the mayor and committee members followed a similar pattern of personal contact, persuasion, assistance in securing financing, and assurance of maximum union cooperation. As can be seen in Table 8.1, from June 1972 to February 1974, JALMC was successful in saving 5 local firms or, in other words, 756 jobs. All these cases involved considerable collective effort and cooperation, but perhaps the most dramatic campaign took place at Jamestown Metal Products. Pooling their resources, 87 of the company's 90 employees raised money to buy the plant from a conglomerate owner planning to liquidate.

In spite of the significant results achieved by their voluntary efforts, JALMC members were keenly aware of their inability to meet the committee's major goals without full-time locally based third-party assistance and leadership. Similarly, Mayor Lundine felt that JALMC needed to forge strong ties with university-based expertise in order to broaden its horizons and strengthen the economic base of the community. After nearly a year of abortive efforts, JALMC won its first technical assistance grant from the Economic Development Administration (EDA) in the spring of 1973. The National Center for Productivity and Quality of Working Life (NCOP) followed with another small grant.

With these grants and funds from the city of Jamestown, JALMC took its first step toward building a professional staff. In June of 1973, James McDonnell took a year's leave of absence from Buffalo State College to serve as JALMC's first coordinator. McDonnell concentrated on persuading local companies and unions to organize their own plant-level labor-management committees (LMCs). At the community level, he organized monthly seminar-dinner meetings to provide a setting for non-adversarial socializing and, also, a forum for new ideas on labor relations, productivity improvement, and the quality of working life. McDonnell also organized JALMC's annual steak fry. An informal gathering for local labor and management leaders, this event has become an occasion for symbolic reaffirmation of the commitments made in 1972.

JALMC's success also brought national recognition during this period. The *New York Times,* the *Wall Street Journal, Business Week,* and *Newsweek* ran prominent articles on JALMC. This helped to

TABLE 8.1
Companies and Jobs Saved as a Result of
JALMC Intervention and Assistance:
1972-1974

Company	Employment at Time of Intervention (Jobs Saved)
Chautauqua Hardware	150
Dahlstrom Manufacturing	150
Jamestown Metal Products	90
Jamestown Metal Manufacturing	166
Jamestown Plywood	200
Total	756

reverse Jamestown's "bad labor town" image and spurred the campaign to attract new industry. In June of 1974, Cummins Engine Company took over the abandoned Art Metal plant, with the expectation that its local employment would reach 1500 in the 1980s. (The prospect now is for employment to reach an eventual 1700.) Cummins officials publicly acknowledged that the improved labor climate growing out of JALMC activities had been a key factor in their decision to locate a major facility to be designed on the basis of self-managing work teams in the Jamestown area.

DEVELOPING IN-PLANT AND COMMUNITY
PROBLEM-SOLVING PROJECTS:
1974-1977

Eric Trist first entered the Jamestown program in May of 1973, but the minimal funds then available through NCOP only provided adequate resources for Trist and his associates to conduct basic diagnostic research and give limited help to a few in-plant LMCs during the start-up phase of their activity. In fact, the development of solid in-plant projects did not become a reality until two years later when JALMC received further funding from EDA, which made it possible to bring two professional consultants, both doctoral students of Eric Trist, John Eldred and Robert Keidel, to Jamestown on a full-time basis.

To build upon the community locus already established for JALMC activity, Trist conceived of an original "two-tier" strategy for a QWL program under the sponsorship of the committee. The aim here was not simply to achieve isolated improvements in individual plants and companies, but, instead, to utilize demonstration pilots in specific cases as building blocks for stimulating broader community change and innovation. To accomplish this goal required the creation of a linking process between local companies and unions such that supportive networks could develop and "multiplier effect" be produced as a result of communitywide learning and the sharing of resources.

To facilitate cross-fertilization between Jamestown area organizations, Trist proposed the development of community QWL working seminars and workshops in which local companies and unions could share with one another their experiences and innovative solutions to industrial problems, as well as learn of developments elsewhere. The beginnings of this type of interorganizational sharing first began in the fall of 1973 when McDonnell organized JALMC's first annual conference. A few of JALMC's budding in-plant committees presented "show and tell" type presentations to representatives attending from other local organizations. Since that time, the annual conference has become an expected yearly event, and through these workshops the community centered strategy originally conceived of by Trist first took hold as a distinctive characteristic of the Jamestown project.

In addition to establishing the community-based strategy during this initial period of involvement, the Wharton team, through their diagnostic research, also helped to catalyze an innovative multiorganizational approach to addressing one of the community's most pressing industrial problems: the impending loss of Jamestown's manufacturing skill base. As employment declined and young workers migrated to other areas, the only skilled individuals left in the community's plants were senior-level employees. In the wood furniture industry this problem had reached the crisis stage, and it was found that the age of the average furniture worker was nearly 55.

Following the lead of a cooperative strategy initially suggested by Trist, McDonnell organized Jamestown's first industrywide training program, linking Jamestown Community College, the United Furniture Workers, and the Jamestown Area Manufacturers Association. Pairs of teachers and skilled craftsmen worked together on a successful program that drew students from many of the area's firms. The classes were held in local factories and financing was provided by the Chautauqua County Industrial Development Agency and the companies.

By the time Eldred and Keidel settled in Jamestown in the late spring of 1975, James Schmatz had succeeded McDonnell as coordinator and JALMC had added an administrative assistant, a young man from the local area, Tony Prinzi.

The new staff members were greeted with considerable reserve. Management was particularly skeptical of what "academics" could do to help Jamestown.

By this time, many of the early LMCs had lapsed into inactivity. Looking back on this period, we now see the problem as one of misguided faith in "good communications" coupled with an inability to develop joint problem-solving.

At first, labor had responded with enthusiasm to the greatly expanded opportunity to discuss problems with management, but generally this developed into a process in which labor was able to identify problems faster than management could solve them. Furthermore, when management did take action, without utilizing the insights that could have been contributed by shop-floor workers, the results often proved to be unsatisfactory to both parties.

This pattern of labor initiative and management response placed both parties in the same reactive or defensive position that they occupied in the grievance procedure. Furthermore, every problem labor identified implied some deficiency in managerial leadership. Managers naturally came to feel frustrated and overburdened by increasing demands for action from labor. In this situation, monthly meetings were often allowed to lapse as managers gave their attention to other matters. In turn, union officials concluded that the meetings had become idle talk, followed by little action, indicating that management was not really sincere about seeking their cooperation.

Eldred, Keidel, and Schmatz worked in twelve companies, seeking to start or revive LMCs. In one case, management abandoned participation before the project was well under way. In three other cases, the union leadership decided to pull out. And in two other cases the consultants withdrew when the projects did not receive sufficient union or management support.

Despite these problems, JALMC achieved some notable successes during this period. For example, Eldred began working with union and management at a Carborundum Corporation ceramic mold foundry as top management was debating the question of whether to make a major investment in redesigning and expanding a very old and inefficient plant (which they had for many years treated as a "cash cow") or to simply relocate and build in another area.

In October 1975, management hired a consulting engineering firm, and, at the same time, the plant labor-management committee formed a subcommittee on plant layout. The subcommittee designed a survey to canvass all workers for ideas and information. Management intended for the subcommittee to collaborate with the consultants.

Paying little attention to the information and ideas provided by the subcommittee, in April 1976, the consultants presented management with a proposal calling for an investment of more than $10.5 million. The plant engineer characterized the proposal this way: "It gave what they thought we wanted but not what we needed.... It was an engineer's dream and a production man's nightmare."

Rejecting the consultant's plan, management now turned the problem over to the subcommittee and company engineers. The survey had brought in 172 ideas, and the subcommittee used 75 percent of these ideas along with the consultant's proposal to work out two alternative plans. In August 1976, the labor-management committee gave these plans to local and divisional management for detailed studies of costs and potential savings.

In April 1977, the president and vice president of the corporation visited Jamestown to tour the plant and review the redesign plans. Labor members of the LMC took the lead in explaining and selling the plan. They regarded it as a victory for labor when top management approved the plan and authorized an expenditure of $5.5 million.

For management, the cooperative project produced a plant judged to be far more efficient than the consultant's plans and at about 50 percent of the cost. For the area, this meant a strengthened tax-base. The groundbreaking in August 1977 became a community ceremonial event, bringing together labor and management officials with city officials, along with former Mayor, now Congressman, Lundine.

During this same period, similar but considerably smaller cooperative, layout restructuring projects were also adopted at two other Jamestown manufacturing operations. At Dahlstrom Manufacturing, one of the five companies saved during JALMC's early years, a mixed team of supervision staff engineers and rank-and-file workers redesigned the layout of a single department. At Hopes Windows Division of Roblin Industries, labor and management jointly designed the entire layout for a new product line.

In 1976, JALMC also broke new ground at American Sterilizer Company, where a labor-management cooperative problem-solving team designed the prototype for a new product, a pilot self-managing work team was established, and a joint approach to product bidding was developed. The product bidding project, which proved to be particularly successful, began when the plant manager was notified that products previously purchased outside could be produced within the company if the local plant could demonstrate its ability to do so at or below the level of outside bids.

The home office enclosed a list of 25 such items. To prepare plans and specifications for manufacture of any of these was a job that would strain the capacities of the plant management staff. Eldred suggested that the manager pick one likely item and have a labor-management task force conduct a study and prepare a formal proposal to the home office.

Temporarily released from their regular jobs to devote full time to the project for several weeks, union members joined the task force with interest and enthusiasm. The result was a detailed plan to produce the item at 15 percent below what AMSCO had been paying. The Jamestown plant had been operating at about $10.5 million annual volume. The new order brought in a guarantee of $1.5 million per year for at least three years and thereby protected some 75 jobs previously slated for layoffs.

JALMC also achieved substantial success in projects at Hopes Windows Division of Roblin Industries. A manufacturer of top quality industrial window frames, Hopes secures much of its business on the basis of competitive bids. When the facts and figures on more complex projects were put together in the traditional fashion by the estimating staff alone, Hopes was successful in only one out of ten cases. Furthermore, management found that the company frequently lost money on the jobs won. With the assistance of Keidel, union and management set up committees of workers, foremen, and engineers to prepare bids. The new system raised Hopes' success ratio to five out of ten. This was not simply because workers revealed to management their true production times. As workers came to understand the elements of costs, they developed new cost-saving ideas. This involvement led them to recognize the importance of cost efficiency in protecting jobs. Earlier, in slack period, workers had slowed down to avoid layoffs.

In this period, JALMC also strengthened its base for facilitating communitywide understanding of QWL concepts and innovations through the Jamestown educational system. An evening QWL course taught by Eldred and Keidel was created at Jamestown Community College, and it succeeded in drawing union and management representatives from both the private and public sectors in the semesters when it was offered. Similarly, at the junior high school level, a teacher, John Sember, worked with John Eldred to develop an eight-week teaching-learning module on industrial relations, productivity, and the quality of working life for junior high school students.

This period also marked the increasing involvement of the author and William Whyte with JALMC under the sponsorship of Cornell University. Cornell had been involved sporadically from the beginning. Robert McKersie, dean of the New York State School of Industrial and

Labor Relations, had consulted with Stanley Lundine and his associates in the early stages and helped them to formulate plans and seek financing. Whyte became interested in late 1975, and Meek began working as a consultant and researcher in Jamestown in January 1976, at first on a half-time basis.

GROWING INSTITUTIONALIZATION:
1977-1981

By mid-1977 JALMC was beginning to show signs of institutionalization. Carborundum's plant redesign and expansion received corporate approval in June. At about the same time, Frank Farrell, president of Hopes Windows, announced that joint labor-management estimating on complex jobs had passed the experimental stage and was now standard operating procedure.

Later in this same year, communitywide skills training in the wood-working industry, which had lapsed into inactivity for nearly two years, was revived and became institutionalized through the formation of a permanent industry-level labor-management committee to assess needs and plan programs. At this writing, over 160 persons or more than 10 percent of the work force in the wood furniture industry have been trained.

Despite such successes, JALMC was in a precarious situation. If the gains already made were to be consolidated and extended JALMC had to move beyond ad hoc responses to in-plant and community problems. Furthermore, from December 1978 into June 1979, JALMC struggled with a financial crisis that threatened its very existence. EDA, which had been the main source of funds, now believed JALMC was established well enough to win support through local and state financing.

To reduce dependence upon staff members recruited from outside Jamestown, JALMC organized a cadre training program in third-party skills for seventeen trainees, nine from unions and eight from management. The program began with a weekend seminar and continued with daylong meetings every two weeks (companies and unions paid for the time off). In plants where they were not employed, trainees worked with staff members on projects. The training program served to recruit and strengthen local talent for the vital third-party role. In October 1979, an alumnus of this program, Richard Walker, became the first graduate to join JALMC as a full-time staff member. Cummins Engine company volunteered to pay his salary for a year as a public service contribution and also to provide a valuable learning experience for a member of its organization. A year later, in December 1980, Carborundum also showed interest in making another cadre graduate, Thomas Matteson, available to assist JALMC on a similar basis.

JALMC was now suffering from the informality and flexibility that had been so advantageous in the beginning. The only part of the committee structure that had been formally specified was that there should be labor and management cochairpersons and a labor-management balance in its total membership.

Furthermore, no procedures had been established for implementing changes in leadership in general or of specific subcommittees, even though leadership positions had the responsiblity and authority to ensure adequate staff performance and the basic economic survival of the committee.

This structure was viable in the early years because staff members, and especially the coordinator, ran JALMC programs, consulting individual members and calling executive board meetings at irregular intervals, using these meetings mainly to report on progress and gain endorsement.

The financial crisis served the constructive purpose of precipitating a reorganization of program and structure. The crisis was accentuated by John Eldred's announcement of his plan to leave the position of coordinator in mid-May.

The executive board chose a former part-time staff member as the new coordinator. John Carney accepted the job on the understanding that he would have the backing of the executive board in important moves to stabilize the program and build for the future.

With the executive board and staff now cooperating closely, JALMC secured from New York State a grant of $95,000 to finance videotape documentaries and other training materials of potential use in other communities. Mayor Steven Carlson pledged the city of Jamestown to continue its support at $35,000. The members of JALMC pledged to raise $35,000 a year from the private sector. Recognizing that it would be impossible for their members to match company contributions, the union leaders pledged an active campaign to raise local money.

The success of JALMC in surviving this financial crisis provided impressive evidence that JALMC was highly valued at home as well as abroad. Management people could have been persuaded earlier to contribute money to the program, but JALMC leaders and staff had always been concerned about maintaining a balance between labor and management. The financial crisis made it clear that JALMC was now established firmly enough with organized labor so that it would not lose credibility with the workers even if companies provided much more financing than unions could provide.

JALMC moved on to reorganize its structure and procedures. A subcommittee wrote and gained acceptance for a constitution and by-laws that, among other things, provided for fixed terms for cochairpersons and labor and management members of the committee. The committee also established a permanent subcommittee on program planning and financing, so that the coordinator and staff members could work closely with the labor-management leadership in both these important fields rather than having the staff carry nearly all the responsibilities.

Staff members have also taken steps to improve the process of third-party intervention in plant situations. Since 1979, staff members have been insisting that any firm that enters into a cooperative project establish such a joint plantwide steering committee and write its own formal charter, including a statement of philosophy, committee goals, and basic ground rules and procedures.

When joint labor-management projects were virtually unknown in the experience of labor and management participants, staff members assumed the role of moderator and discussion leader at the meetings of project committees. Although this may have been necessary in the early stages, it had the disadvantage of maintaining the parties' dependence upon the staff person.

Now, although staff members may agree to conduct the first meetings in which the parties are considering whether or not they want to establish a plant labor-management committee, they proceed only when both parties agree to jointly take over meeting leadership and project coordination themselves. The practice now is to have labor and management cochairpersons, at all levels, who alternate between taking minutes and chairing each meeting.

This approach first began to develop at Corry-Jamestown Company after a bitter nine-week strike in 1976. The plant, employing 500 workers, manufactures high-quality metal office furniture and is located in Corry, Pennsylvania, a small city approximately 25 miles southwest of Jamestown.

Because of the company's distance from Jamestown, JALMC coordinator, Jim Schmatz, devised a plan to sit in on committee meetings on a limited basis. This resulted in the formation of a joint steering committee that spun off various problem-solving projects to be tackled by subcommittees. The steering body and subcommittees were chaired and coordinated by plant management. The Corry-Jamestown committee carried out several successful projects, but the first major undertaking involved the introduction of a productivity-gains sharing program. For this project, management brought in an outside consultant with special knowledge of productivity gain-sharing plans. By this time, consultants Chris Meek and Patrick McGinity of JALMC, had assumed the third-party role at Corry-Jamestown. They worked closely with labor and management to structure the process of introducing the plan so as to make sure it was jointly implemented and that all workers had full opportunity to express their concerns and interests, as well as gain a basic understanding of the gain-sharing plan that was adopted.

Management estimates that, compared with preplan figures, productivity was up by nearly 15 percent in 1978, 22 percent in 1979, and approximately 20 percent in 1980. This meant worker bonuses of 7.5 percent of wages in 1978, 11 percent in 1979, and 15 percent in 1980.

Encouraged by the success of this joint effort, management and labor have worked together to develop a program in which worker participation in problem solving has become fully integrated into the process of managing the plant.

This period of institutionalization has also been marked by further strengthening and elaboration of the interorganizational learning process. In the past, the technical assistance required for the transference of innovative problem-solving approaches to new sites was primarily provided by JALMC staff consultants, but today we find direct "plant-to-plant" consultation and teaching to be an approach used with increasing frequency. One of the first cases of such direct assistance took place in February 1979 when Corry-Jamestown management consulted with Carborundum management on the applicability of a productivity gain-sharing plan to their operation. Later, that same year, labor and management members of the Carborundum redesign subcommittee provided basic training in cooperative plant restructuring methods to a similar group at Corry-Jamestown.

As in the case of Carborundum and Corry-Jamestown, a similar pattern of direct exchange and mutual assistance has developed in a number of other instances.

CONCLUSIONS

The results of such a far-reaching program as JALMC defy any brief summarization, but it is possible to examine several important indicators.

In the initial crisis period during which the committee was organized, JALMC was directly responsible for the immediate saving of 756 jobs. Since that time, the committee also succeeded in directly averting the closure of another plant, Watsons. It also helped to ensure the survival, retention, and growth of many other companies through the development of plant-level cooperation and problem solving.

Beyond this, JALMC is generally credited with reducing the incidence of labor-management conflict by changing the labor relations climate.

This is truly a difficult point to prove, but whatever the reason it is clear that manufacturing strikes decreased in Jamestown from 39 to 31 when the decades of the 1960s and 1970s are compared and, even more dramatically, the number of strike days was reduced from 1442 to 749. Despite the difficulties involved in proving or disproving that JALMC helped to reduce work stoppages in Jamestown, the committee's efforts clearly did succeed in changing Jamestown's image as a "bad labor town."

In sum, these accomplishments reveal some impressive gains in terms of both manufacturing employment stability and growth. A sample of what some of these achievements add up to in terms of employment are illustrated in Table 8.2. As can be seen, the several JALMC efforts cited in the table (which by no means are all inclusive) indicate that committee activity has resulted in a direct saving of 1708 manufacturing jobs. Since its inception, it has helped to ensure and facilitate the creation of 2501 new jobs and resulted in a total employment impact of 4164 manufacturing jobs saved and created. Clearly, this is an impressive record considering that the majority of local companies represented in this table would have either folded or relocated to another part of the United States. Similarly, it is highly unlikely that Cummins Engine would have decided to build a new operation in Jamestown, especially one based completely upon shop floor self-management, had the city not successfully changed its negative image. Finally, it should be noted that such achievements were made with a relatively modest investment on the part of the federal government. David Stockman, for example, cited a figure of $60,000 as representing the average cost per job created through Economic Development Administration technical assistance during his early advances on cutting the federal budget. In contrast to this clearly exorbitant expense, the cost to EDA in terms of just the jobs that were saved in Table 8.1 was only $275.18 per job. The jobs that were created cost $186.92 per job, and the total cost of EDA assistance per job with respect to the overall employment impact was only $112.87.

JALMC has also made another important breakthrough considering the variety of organizations aided by this group. Most well-known

TABLE 8.2
Sample of Employment Protected and Increased through
JALMC Intervention and Assistance: 1972-1981

Company	Nature of Assistance	Jobs Protected	New Jobs Created	Total Employment Impact: 1981
Chautauqua Hardware	plant saving	150	265	415
Dahlstrom Manufacturing	plant saving	150	200	350
Jamestown Metal Products	plant saving	90	20	110
Jamestown Metal Manufacturing	plant saving	166	169	335
Jamestown Plywood	plant saving	200	0	155[a]
Watson-Afro-Lecon	plant saving	68	47	115
Cummins Engine	new industry development	300	1,700	1,700
Carborundum	plant redesign	300		300
Falconer Glass Industries	five-year contract	359	100	459
AMSCO	subcontract bidding	75		75
Hopes Windows	subcontract bidding	150		150
Total		2,008	2,501	4,164
Cost of EDA Technical Assistance per Job		$275.18	$187.92	$112.87

a. Jamestown Plywood actually reduced its overall work force in 1975 by 45 because of previous overstaffing. Employment has remained stable since.

projects in productivity and quality of working life have been carried out in companies having both money and staff to spare. Although in some cases major companies (Carborundum for example) have been involved with JALMC, much of the activity has been carried out in small firms or plants, some of them in precarious financial condition. JALMC could not survive as a luxury to be dispensed with whenever money was short.

Thus JALMC has broadened the methodology of planned organization change and offered an alternative strategy for merging both the concern for improving the quality of working life and the need to revitalize economically the deteriorating industrial communities of the United States. Furthermore, by working from the strategic perspective of the community, JALMC has pointed the way to solving consultation problems that many small companies face.

A large company generally has its own staff of internal consultants. Furthermore, if outside consultants are needed, the internal staff may have sufficient experience and expertise to distinguish effective technical assistance from mere showmanship. The small firm often lacks not only internal staff resources but also the experience and ability to find consultants whose performance will be worth their cost.

The lesson in this case is not simply to avoid out-of-town consultants. As noted above, at Corry-Jamestown an outside consultant was used effectively. However, to the extent that the staff of JALMC and its in-plant committees can develop their own skills, strategies, and methods for solving common local industrial problems, they can serve as consulting resources to Jamestown area firms and unions needing help, and on essentially a no-cost basis. Of course, the actual previous performance of these resources can easily be checked with other local labor and management people, and if, in fact, outside consultation does prove necessary, JALMC can work with union and management to locate the right consultant and fit him or her into the local scene.

The community-based nature of JALMC also offers advantages in starting and terminating projects. In the usual situation, if management decides to end a project, the consultant leaves town. At a later time, if union and management decide to resume work on a joint project, they may not be able to find consultants who have had previous experience with their organization. In the Jamestown case, terminations or failures are not so likely to be permanent. For example, in the cases of AMSCO, Dahlstroms, and Falconer Glass Industries, some months after projects with JALMC were broken off, management and union leaders agreed to invite JALMC to resume activities.

Altogether, the JALMC efforts provide a prototype of an American approach to genuine industrial democracy. The benefits of such a community-based program extend far beyond the gates of the factories. Jamestown has been building its community organizing capacity. The joint projects carried out within plants and in interorganizational cooperation have been providing valuable social learning. The successes achieved have reinforced the commitment of the participants to this development strategy and have increased their confidence in Jamestown's ability to solve community problems. Through JALMC and its related activities, in less than a decade, Jamestown has changed from a dying industrial city to an exciting place in which to live and work.

REFERENCES

LUNDINE, S. (1978) Personal interview (January).
McMAHON, H. G. (1958) Chautauqua County: A History. Buffalo, NY: Henry Stewart.

9

LABOR-MANAGEMENT STRUCTURES
IN THE LARGE CITY

Robert W. Ahern

This chapter will concentrate on the functions that an area LMC can undertake in a large metropolitan area through an examination of the experience of the Buffalo-Erie County Labor-Management Council (BECLMC).

Buffalo, New York, like many other northeastern cities, has been afflicted with a steady erosion of its manufacturing base as firms have decided to close local facilities and/or relocate their plants to other areas. The deleterious effects of this process have included a dwindling tax base, a negative ripple effect throughout the secondary and tertiary industries, and, most important, the direct loss of thousands of high-paying jobs. Employers give many reasons for closing plants: high taxes, shifts in markets, obsolete plants and equipment, high wage rates, poor productivity, and, sometimes, poor labor relations. The temptation of the "greenfield" site, in which new equipment can be laid out in a "union-free environment," becomes very strong when corporate planners are considering new capital investment decisions; whatever the reason for these decisions, deterioration of the manufacturing base was a serious problem.

To respond to this advancing problem, in May 1974 the mayor and the county executive created a committee for economic development. This committee began to bring together the various agencies that had been working on economic development issues, often at cross-purposes, to develop a unified plan for plant retention, expansion, and industrial recruitment. Organized labor was made part of this effort.

As the committee focused on the plant retention problem, it became clear, as first BECLMC Co-Chairman Daniel A. Roblin explained, that "while leaders in Buffalo could not directly or immediately affect the high tax rates, the marketing strategies of outside firms or their capital investment decisions, those leaders could do something, directly and immediately, about labor relations in the Buffalo area." The response of labor was prompt and positive: The Buffalo AFL-CIO passed a resolution instructing its president to seek out management leaders and attempt to put together a construct for communitywide labor-management cooperation. Their motivation for action was obvious, for, as George Wessel, president of the Buffalo Area AFL-CIO Council and BECLMC co-chairman, put it, "The name of the game is jobs!"

At this point the example of the Jamestown Area Labor-Management Committee (JALMC) drew the committee's interest. JALMC had directly attacked the problem of plant retention by establishing in-plant labor-management committees (LMCs). These committees had improved plant-level labor relations and they were beginning to show an ability to improve productivity and quality. Thus BECLMC had before it an approach that would focus on several elements important to retaining local plants: labor relations, productivity, and quality. Furthermore, if the increases in productivity and quality were sufficient, the reduction in unit labor costs could make Buffalo's relatively high wage-rates more competitive. James McDonnell, JALMC's first coordinator, consulted with the committee. He explained the general concept and the details of how in-plant LMCs worked.

Roblin and Wessel then put together a working group of five labor and five management officials to meet with the mayor, the county executive, and the congressional delegation. Several meetings were held in mid- and late 1975 in which the decision was made to establish an areawide LMC for the Buffalo area. The county appropriated $30,000 to fund the effort and I was hired to be staff director on December 1, 1975.

One of BECLMC's first staff studies concentrated on the local labor-relations climate. Economic development people reported that Buffalo had a terrible reputation as a "bad labor town," and that this image was a major obstacle to industrial recruitment. The co-chairmen felt that this characterization of Buffalo was exaggerated and outdated, but still they recognized the problem had to be faced head on.

To determine just how close Buffalo's image was or was not to the real situation, the staff conducted a study of time lost due to work stoppages in 36 major cities across the country. The results made it clear that Buffalo had a terribly serious problem: The myth was a reality. For example, in 1970 the Buffalo standard metropolitan statistical area (SMSA), which includes both Erie and Niagara Counties, lost 1.05 percent of its working time due to strikes. This record ranked it the third most strike-prone city in the nation; only Kansas City with 2.58 and Detroit with 1.26 percent were higher. In 1971, Buffalo moved up to second place, and in 1972 it was first in the nation. In fact, the staff discovered that four times in the six-year period between 1970 and 1975 Buffalo was first, second, or third in working time lost due to work stoppages. Thus as the BECLMC began to take shape in late 1975 and early 1976, it was clearly not facing a hypothetical problem. In comparison to the rest of the country, Buffalo's labor relations were very poor.

STRATEGY FOR ORGANIZING THE COUNCIL

Because basic funding was in place and a staff director had been hired before the council's formation, the director and the co-chairmen were able to work together in laying out the necessary groundwork and strategy for a successful effort. As they met and discussed priorities and concerns, the following principles emerged as key guidelines for establishing the Buffalo-Erie County Labor-Management Council.

(1) Council members would all be private-sector people; no government personnel or public-sector leaders would be directly involved.

The rationale for this principle was fourfold. First, the primary problem of work stoppages and the erosion of the manufacturing base were private sector problems. Second, given the yeasty political picture in Buffalo, with a Democratic mayor and a Republican county executive and a less than tame city council and county legislatures, it was felt that the council should maintain a strictly nonpartisan stance. Third, the opinion was that public-sector labor relations were still too immature and wedded to the adversarial system to be able to explore cooperation effectively. Finally, it was clear that the magnitude and urgency of private-sector labor relations problems were such that the workload would simply preclude public sector activity.

(2) Only the chief operating officials of the companies and unions in the area would be members.

This guideline, which also has become a rule of thumb for in-plant LMCs, flowed from two central premises: (1) If the chief official of the respective organization is not involved, the power of that organization is not involved; (2) if the chief official is involved, a commitment to the program is communicated to the other side and to the entire community.

(3) Management members would be chosen from Buffalo-based firms and large branch plants.

It was believed that chief operating officers of firms with headquarters in the Buffalo area would most identify with the community and have the greatest vested interest in its development. However, the co-chairmen recognized that the top managers of some of the large branch plants were powerful members of the management community and should be part of the process. The operative principle was to involve enough key managers so that the power structure of the entire management community could be drawn upon when necessary.

(4) The entire labor power structure would be drawn into membership.

On a local level it is rare for all international or local unions to be members of the same central body. Further, there are other coalitions, formed for specific purposes, that have significant power, and that sometimes support and in other cases counter the local AFL-CIO council. For instance, Buffalo has a Maritime Trades Council (MTC), which has often made political endorsements directly opposed to those of the AFL-CIO Council, even though many members of the MTC also belonged to the Council.

Thus an in-depth look at the union power structure made it clear that in addition to charter members, the IUE, the Carpenters, the district director of the Steelworkers, the area director of the UAW, and the international vice presidents of the Grainmillers and the Longshoremen's Associations would also have to be recruited if BECLMC wanted to utilize fully the power of labor in the area.

INITIAL DECISIONS

During December 1975 and January 1976, the co-chairmen and the executive director discussed the council with several prospective members and recruited four additional members on each side to ensure solid backing from both the management and union power structures. Thus when the council formally announced its creation on February 12, 1976, there were nine members on each side.

An advisory committee was also formed. This committee was established to provide BECLMC staff with technical advice on both labor relations and human resources management. It was also felt that committee members could be helpful in tapping the informal network of contacts among industrial relations experts throughout the area. Accordingly, three vice-presidents of industrial relations from management, three union staffers, three local Federal Mediation and Conciliation Service (FMCS) commissioners, and the head of the New York State Mediation Board were appointed to advise the council.

During its first few monthly meetings, the council made a series of decisions that have since played a central role in shaping the council's role and development. Some of the most important are described below.

(1) The council decided to concentrate on the private sector manufacturing base, and not become involved in the public sector.
(2) BECLMC's attorney was directed to incorporate the council as a not-for-profit corporation.
(3) The council would keep a very low profile with the media.
(4) The council would seek out smaller plants where the probability of improving labor relations through in-plant LMCs would be high.
(5) The council would involve itself in backstage efforts to end long-term strikes.
(6) Because of limited resources, council staff would provide third-party services to in-plant LMCs for only a limited period from nine months to one year.
(7) The local economic development effort would have to be restructured.

When the council began its work in 1975, the economic development fraternity for the Buffalo area was divided and working at cross-purposes, even though the local economy was in severe decline. To rectify this situation, the co-chairmen and a distinguished member of the advisory committee, Harlan Swift, Jr., chairman of the board of a local bank, devised a scheme for bringing together the warring parties under the structure of the local industrial development agency (IDA).

In New York State, IDAs are creatures of the state legislature established to provide industrial revenue bonds and other financial vehicles that encourage development. Buffalo's IDA had been tucked away under the guidance of the Chamber of Commerce and its vice president for economic development. The scheme developed by BECLMC leaders was to expand the board of the IDA to include the heads of all economic development agencies in the community so that any non-cooperating agency head or agency could be brought in to account for themselves before the board. In this way it was felt that economic planning and development could proceed under the guidance of this expanded board and be independent of all cooperating agencies. It could also seek to constantly expand its functions to fill in gaps in the development framework. The decision was made to name the members of the IDA board by their office. This required enabling legislation from the state to create this unique organization. An intense lobbying effort followed and the legislation was passed. When it reached the governor, his legal staff raised constitutional questions about naming the individuals on the board by office rather than as individuals, and recommended to the governor that the legislation be vetoed. An even more intense lobbying effort convinced him to sign the legislation.

The council had, as its specific objective, the interlocking of the board of directors of the IDA and the council. BECLMC labor and management co-chairmen became the chairman and vice chairman for the IDA, respectively, and two other council members were named to the new board. The council had therefore become a moving party and a vital part of the unified economic development effort. At the time it was hoped that all the agencies could be housed in a "one-stop" shopping center. Therefore, for the first few years, BECLMC maintained its offices in the same building as the IDA. But after a few years its joint location led to a perception that the council was an arm of the IDA, and the council made a decision to move to separate headquarters.

IN-PLANT LMCs

The major function of BECLMC is to create a network of plant-level LMCs that address the problem of labor relations climate, productivity, quality, and job satisfaction through employee participation in workplace decisions. This is the heart of the council's efforts, and the staff devotes the greater portion of their time and energy to developing and sustaining in-plant LMCs by providing a broad range of third-party process consulting services. These services include the following:

- exploring the LMC process with the parties to determine readiness, depth of understanding, and commitment;
- briefing all levels of labor and management leadership;
- attending LMC meetings to make sure the parties properly implement the mechanics of the process;
- advising the parties on process philosophy, goals, and programs;
- providing expertise in organizational change, human resource planning, and productivity improvement;
- counseling individual leaders;
- training participants as required;
- auditing the process and recommending action; and
- coaching internal process consultants.

Because of the council's limited resources, one of its early decisions was to limit staff consulting services to any one organization to a period of nine months to one year. Field experience soon showed this to be an unrealistic decision. Organizational change takes much longer. The present guideline is eighteen months to two years. Beyond this point, there is an ongoing maintenance function that includes the items below:

(1) monthly audit of LMC minutes;
(2) periodic field audits as requested;
(3) orientation of new players;
(4) regeneration sessions to stimulate the parties and the process; and
(5) troubleshooting as requested.

Today, BECLMC is probably the largest network of in-plant LMCs in the country. They are well established in all sizes of organizations and in all types of industries, including, for example, a gas utility with some 3000 employees, a newspaper with more than 1000 employees, a bus company with 1200 employees, a major textile manufacturer with 1500 employees, a major tire company with 1200 employees, and dozens of committees in a variety of smaller firms. Thus the in-plant LMC has demonstrated its effectiveness as a vehicle for change in a wide range of environments, and it has also shown that such change can become institutionalized. Many of the council's LMCs have been in existence for almost ten years and have passed through two or three negotiations. Thus it appears they are becoming a permanent part of the way in which unions and management do business with each other.

THE SATELLITE AREA LMC

When BECLMC was first getting off the ground, many experts were skeptical about the possibility of a large city such as Buffalo succeeding because of its size and diverse industrial and union population. Earlier committees, such as JALMC, had in part benefited from the fact that smaller communities, such as Jamestown, tend to have clearer, more cohesive, and relatively homogeneous labor and management interest groups. In contrast, the Buffalo area is made up of a number of very distinct communities whose interests do not necessarily overlap and that have their own special labor and economic development climates.

To solve this problem and meet the specific needs of Buffalo's several labor-management communities, the council developed the Satellite Area LMC. These LMCs are composed of union and management leaders from a particular geographic area. They permit broader participation in the work of the council and a clear focus on the problems of that specific community. Council staff provide the organizational and operational legwork for these groups as well as ongoing guidance. An excellent example of this concept is one satellite committee that BECLMC helped to establish on Buffalo's waterfront.

The waterfront LMC was one of the council's first efforts. Labor relations in the port had been explosive and bitter for many years. Not only were the Grainmillers and the International Longshoremen's Associations militant and tough, but relationships between the two were extremely strained. During the early 1950s, a strike erupted into a major riot when one union crossed the other's picket line. In early 1976, the parties came together and established a committee the purpose of which was to improve labor relations in the port area. This group consisted of the plant managers in the grainmilling industry (General Mills, Peavey, International Multifoods, Pillsbury, Seaboard Allied), representatives of Local 36 of the Grainmillers International and ILA Locals 128 and 928. This committee met monthly, and over the years achieved the following objectives:

(1) A substantial improvement in port-area labor relations.
(2) Major revision of the labor agreement between the cargo industry and ILA 128. The revisions greatly increased the flexibility of human resources and therefore made the stevedore companies more competitive in bidding for business on the Great Lakes.
(3) The council conducted a joint study of work practices in the grainmilling industry with the companies and ILA 928. This study resulted in a focus

on difficult work practices that over time the parties have modified through negotiations to improve flexibility and productivity.
(4) In 1978 this area LMC guided a comprehensive study of Buffalo's port facilities. The Economic Development Administration funded the study, which resulted in several important cost-saving recommendations.

One of the most critical recommendations that resulted from the 1978 study was that monies be allocated to obtain one heavy-lift crane (230-ton capacity) and a heavy-duty bulk cargo system. Through two local state assemblymen, legislation was introduced to provide $1.6 million for this equipment. The effort was successful, and the funds were included in the FY 1980 supplementary budget. The equipment was put in place for the 1981 shipping season, and it has since helped to significantly increase business due to expanded capability.

The 1978 study also resulted in many other important improvements such as the following:

(1) the establishment of an office of waterfront development through the Erie County Industrial Development Agency, which has since grown to become the Niagara Frontier Transportation Council
(2) an in-depth comparative analysis of labor costs and practices in other Great Lake ports
(3) a thorough analysis of the feasibility of straightening the Buffalo River by the city and the Army Corps of Engineers
(4) the development of two in-plant LMCs in the grainmilling industry

By 1980, labor relations on the ports were so good that the president of Local 128 of the ILA was often involved in marketing conferences with prospective shippers and steamship lines. In fact, he accompanied the president of the major stevedoring company on several marketing trips around the Great Lakes.

A second illustration of successful use of the satellite area LMC concept is provided by the community of the Tonawandas (the cities of Tonawanda and North Tonawanda), which is located on the Niagara River almost equidistant between Buffalo and Niagara Falls. It is the western terminus of the Erie Barge Canal, which serves to separate the two cities and also Erie from Niagara County. The total population of the two cities is 54,453. The Tonawandas Area LMC came into being in 1981. It is structured like BECLMC and performs similar functions. Since 1981 this LMC has accomplished the following:

(1) Fostered the creation of seven in-plant committees.
(2) Conducted ten monthly meetings a year, concentrating on raising the consciousness of local labor leaders concerning labor-management

cooperation. Speakers in the last two years have included the county executive, Congressmen John Lafalce and Stan Lundine, Sam Camens of the United Steelworkers of America (USW), and other experts in labor-management cooperation.

(3) Instituted a series of periodic meetings at which labor and management leaders and in-plant facilitators meet separately to discuss the real problems and opportunities of the labor-management cooperation.

(4) Implemented a training program for middle management that is designed to meet the special sensitivities of this group when the in-plant LMC process is introduced.

(5) With the cooperation of the city council, staff intervened in one major and two minor labor disputes (two of which are described below).

(6) Effectively lobbied local legislators to get special tax considerations for marginal companies in the area.

The Tonawandas Area LMC has proven that the satellite concept can be used to increase the network of in-plant LMCs in an area and at the same time focus local energies on local problems. In contrast to BECLMC, the Tonawandas LMC meets social as well as educational needs at its monthly meetings. The formal meeting provides education and dialogue, and the informal cocktail and dinner hours have furnished a real opportunity for those involved to get to know one another as people. This has had beneficial effects for the entire process.

In the future the council will consider spreading the satellite area concept to other communities throughout Erie and Niagara Counties.

INTERVENTION IN LONG-TERM
LABOR DISPUTES

The council intervenes in long-term labor disputes that have severe effects on the community and its labor relations image. The working definition of a "long-term" dispute is a strike of twelve weeks or more. No intervention is made without coordination and cooperation with the mediating agency or agencies. Such interventions are very delicate situations and can perhaps best be defined by describing two actual cases.

The first case took place in 1977, when a raw materials manufacturer that had a long history of poor labor relations had a contract dispute that led to a long strike, which eventually resulted in the hiring of economic replacements. Picket-line violence ensued and there were legal actions against the union's leaders. The council intervened, again with full knowledge and cooperation of the federal mediators involved, met with the respective parties, isolated the key issues, and then created a scenario for settlement. Congressman John Lafalce called a meeting of the parties and the federal mediators, and with a diligent all-night effort was able to achieve a settlement. In this case the council followed up by establishing an in-plant LMC, which in one year and a half had turned the labor relations climate completely around. Prior to the LMC, roughly 500 grievances were filed, and 30 to 50 were up for arbitration each year. In the three years following the establishment of the LMC, only three grievances were filed. As a result of the change in the relationship, this company, a few years later, announced a major plant expansion.

A second interesting case involved a long-term strike between a local metal fabricator and the steelworkers. The company, after a series of negative ratification votes, gave notice to the union and the public that if the tentative agreement it had reached with the union was not ratified, it would close the plant.

In the face of another negative vote the company made a public announcement that it was indeed going to close. At this point the council intervened. Through the management co-chairman a meeting was arranged with top-level executives in the company's New York office and because of the executive director's long-standing relationship with the district director of the steelworkers, a meeting was arranged with top union officials in Atlantic City, where the union was in convention. These two meetings were held on successive nights. Later through phone conversations with the international vice-president responsible for negotiations and top management of the company a game plan was developed that was executed when the parties met to bargain the effects of closing the plant. This plan was effectively executed, and this plant, which the public thought was doomed, is still a functioning and healthy part of the Buffalo community.

Once the council learned this very delicate drill, it was able to intervene in several other long-term labor disputes in an effective manner. However, with the dramatic decline in work stoppages that resulted from BECLMC's efforts, this skill has atrophied somewhat. It has therefore been put to use in areas other than manufacturing. In 1983, the council helped end a very bitter nurses' strike at one of the local hospitals that had been going on for more than 12 weeks.

HUMAN RESOURCES MANAGEMENT
AND INDUSTRIAL RETRAINING

In August 1977, Bethlehem Steel announced it was reducing the work force at its Lackawanna plant from 11,500 to 8,500. This reduction devastated the community. The Economic Development Administration (EDA), knowledgeable of BECLMC's human-resources efforts, asked the council to do something for the laid-off employees. The council accepted the challenge by setting up what came to be called the "transition center." Located about a mile from the plant, the center housed all community support agencies that could aid the displaced workers. Originally, the concept was simply to centralize these services in this "crisis center." However, it soon became apparent that the real need was for outplacement in jobs or well-designed training programs. The center then took as its major objective the job transitioning of these laid-off workers to other employment, hence the name "transition center."

The center was a joint effort of the council, the New York State Department of Labor (DOL), and a host of community agencies that included both the city and county CETA organizations. Council staff designed and built a unique data base, analyzed labor market demand, brokered the development of training programs, ran a highly successful outplacement office, and managed the entire operation. State DOL personnel accepted administrative direction from the council; they tested, counseled, and placed center registrants in training programs. They also used the wide network of DOL offices to place registrants.

Within two months, the mechanisms were in place to discover labor demand and to create and fund training programs to meet that demand. Programs created included those for skilled welders, precision machine operators, industrial electricians, drafters, industrial maintenance persons, and tool-and-die apprentices. Registrants were also placed individually and in groups in established training programs that could demonstrate high placement rates. One enterprising registrant used trade readjustment act (TRA) benefits to go to Las Vegas to learn to repair gambling machines; he then found a job in Atlantic City.

The center ran for eleven months. In that time, 1891 unemployed steelworkers registered at the center. Of these, 1231 were at work or in training programs when the center closed: an effectiveness rate of 65 percent. The rate would have been higher but many of the 660 participants refused training programs in favor of waiting to be recalled to their jobs at Bethlehem.

Four years later, December 1982, Bethlehem announced it was closing most of its operations in Lackawanna. This meant that another 7,000 employees would be out of work. The chairman of the county legislature, an ex-officio member of the council, and the county executive asked BECLMC to set up countywide transition centers, not only for the Bethlehem workers but for all the unemployed of western New York. The council took this assignment with the stipulation that, after establishing its mechanisms and procedures, it would turn the operation of the centers over to a governing board that would be established for that purpose. From March through October 1983, council staff ran the centers. At that point, the centers became independent of the council except for their fiduciary control, which was relinquished on May 1, 1984.

The council feels that major league human resource programs such as running transition centers for thousands of workers are outside its mandate and scope. Its experience in 1977-1978, and again in 1983 made it clear that the managerial burden of such efforts detracts from the main work of the council.

RESULTS

The foregoing discussion describes only a part of the council's activities. In addition to those areas described above, BECLMC also provides a wide range of labor relations and labor-management training programs, and for all cases in which an in-plant LMC is organized, each party is asked to designate a leader to be trained as an internal process consultant. Such individuals watch over the LMC process when council staff leaves the site, and the committee moves into the maintenance stage. The council also sponsors an annual conference to raise the consciousness of the labor and management community concerning the effectiveness of cooperation. Typically, panels of labor and management co-chairmen from ongoing LMCs present the history of their committees and describe specific results. They then engage in an open dialogue with seminar participants. Panels are chosen from in-plant committees that show that process is effective in large as well as small organizations and across a broad variety of industries. Finally, it should be noted that the council has developed the innovative idea of labor and management network meetings. In 1983, the council began to hold separate meetings for the union and management leaders, as well as the growing number of process consultants so that each side and the facilitator neutrals could learn from their counterparts in other organizations through the exchange of information and experiences.

All these programs and activities have combined to help yield some significant labor relations and economic gains for the greater Buffalo area. Certainly, one of the most dramatic indicators of BECLMC's accomplishments is the dramatic change in the community's labor relations climate that has developed since the council's formation in 1975. Figure 9.1 helps to tell this story and to illustrate what the labor relations reality is today.

In the seven years since the council was founded the strike rate in Buffalo has decreased by 90 percent. Buffalo SMSA, with a 1982 strike rate of .04 percent had the lowest rate of all the SMSAs in the country that have 25 percent of their work force unionized. This record was continued in 1983 and 1984. The formation of cooperative efforts in previously high-conflict organizations has helped considerably in bringing about this change of affairs. BECLMC has established LMCs in several firms that in the past had chronic histories of poor labor relations and strikes. Several of these had back-to-back strikes in their past two negotiations prior to the establishment of an LMC. These plants now have excellent labor relations, a cooperative employee participation program, and they have had no strikes even though they have been through two or more negotiations. Indeed, with one exception, negotiations went more smoothly than they ever had, despite very trying economic times.

Clearly in-plant committees are the area in which BECLMC has shown some of its most impressive achievements. The council currently provides ongoing service to some 74 LMCs in 21 different organizations; 30,000 hourly and salaried workers are directly involved. Each year approximately 8 new organizations establish in-plant committee networks under the council's guidance.

It is not possible to detail the results in all these LMCs, but it is possible to spell out the general results that occur in all LMCs. To illustrate the process, the history of a few in-plant committees that have advanced into the structured problem-solving stage will then be discussed. All committees generally report the following:

- improved communications;
- a better labor relations and employee relations climate;
- reduction in grievance activity;

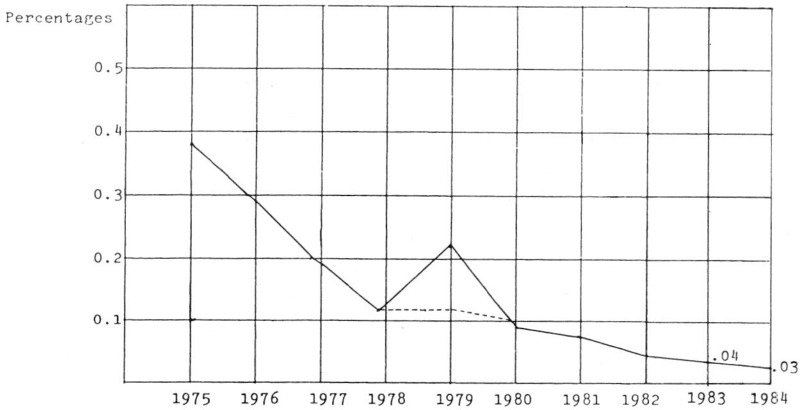

NOTE: The dotted line shows the strike rate without the nationwide Westinghouse strike.

Figure 9.1   Time Lost Due to Work Stoppages: Buffalo 1975-1984

- substantial employee involvement in workplace decisions;
- growth in trust and understanding between the two parties;
- faster response time on decisions affecting the workplace;
- improved physical environment;
- rapid and broad development of leadership skills on the part of both union and management leaders; and
- discovery of people with leadership potential.

Those organizations that have good measurement systems and committees that have set clear performance goals generally report the following results:

- productivity (general order of increase 20-25 percent);
- improved product quality and reduced waste;
- reduced absenteeism (general order of decrease 50 percent);
- reduction of grievances to all-time low levels;
- professional and objective negotiations; and
- increased job satisfaction, motivation, and morale at all organizational levels.

Clearly, such results indicate that the LMC is an important technique for change in unionized firms. However, proceeding into the structural problem-solving stage of the process requires that the parties have an explicit understanding concerning productivity, job security, and the sharing of gains resulting from cooperation. An absence of agreement on these important considerations will make it difficult, if not impossible for the parties to reach the advanced stages of the process. For those that do reach these stages, such as the two cases cited below, the rewards can be well worth the effort.

The tire industry has been in trouble for many years. In 1977, management and local 135 of the United Rubber Workers at Dunlop Tire started a cooperative effort to improve plant performance and make the company more competitive. Almost a hundred meetings were held, and during the next few years there was a dialogue concerning the plight of the company and what might be done about it. By 1980, the plant's performance was still uncompetitive and the firm reluctantly had to consider closing. The union, with more than a third of its members laid off and facing a possible plant closure, agreed to a $1.05 wage reduction package in exchange for a plantwide productivity bonus and the installation of an LMC process. Productivity increases needed were on the order of 25 to 30 percent.

The LMC began in late 1980, and two years later productivity and quality had improved (approximately 25 percent) to a point that the plant became profitable. Laid-off workers were recalled, and the company was one of the very few in Buffalo that was hiring.

The council assisted this LMC for about eighteen months. During that period the parties developed enough expertise in employee participation not only to go about it alone but to also develop entirely new and broad-ranging employee involvement programs under the aegis of its steering committee.

Since 1982, performance has continued to improve. New targets have been set and met, and, in January 1984, three years after contemplating a plant closure, Dunlop announced plans to build a new $150 million tire line in Buffalo.

Similarly impressive results have been achieved by Exolon, an abrasive manufacturer and Local 1411 USW. Two years ago the company threatened to close its plant down after two consecutive years of losses.

According to the director of operations and LMC management co-chairman, "The LMC has been the major factor in keeping the plant alive . . . there are now a couple of hundred heads working on a problem, rather than one." The union president explains, "We were on borrowed time, but we survived."

Since the formation of an LMC, productivity has increased and by a figure of 16 percent in 1983. Grievances declined from thirty a year to zero, and the flow of red ink has turned to black.

One major accomplishment of the LMC was a shift from bonuses based on an incentive system to profit sharing. Under the old system, the company was forced to borrow money to meet its incentive pay obligations. The LMC designed and now administers a profit-sharing plan that has proved to be a true motivator.

In 1984, Exolon worked full schedules at full capacity with an additional 10 percent increase in productivity. The employees accepted that increase with the response, "We'll do whatever we have to do to achieve it." Consequently, Exolon is at full employment, is even hiring new personnel, and paid out a profit-sharing bonus in early 1985.

FUNDING AND INSTITUTIONALIZING
THE AREA LMC

Perhaps the most impressive sign of BECLMC's success is the extent to which the community itself has adopted the effort and made the council locally self-sufficient. Many area LMCs have failed to reach this level of maturity, and as a result they have typically died out once state or federal funds have dried up.

The Buffalo story is unique because BECLMC has never received one federal or state dollar to support its labor relations objectives despite the Council's and Congressman Jack Kemp's active roles in bringing about the passage of the Labor-Management Cooperation Act of 1978. Federal funding has only been received for the human resources functions performed by the council that were described earlier.

In 1975, the county legislature appropriated $30,000 as seed money. Over the years, appropriations from the city and private-sector donations from union and management organizations have increased steadily. Today, the county's appropriation of $130,000 supports a core staff of three, and this budget is supplemented by donations from the private sector, payment for council-provided training, and, most recently, consulting activities. To date, these supplementary sources of funding have varied from $3,000 to $30,000 year. It is anticipated that the budget will grow significantly as the council expands the range of training programs it provides and accepts more consulting assignments.

Early on, the council decided that staff would limit the length of service to in-plant LMCs. The first time-frame of nine months proved naïve, but the premise of making LMCs self-sufficient was sound. As a result, a three-person staff has been able to service a current network of 74 LMCs in 21 firms and two satellite LMCs, and start a minimum of eight new LMCs a year. With a county cost of $130,000 for this effort, which is only .02 percent of the total 1984 county budget, our members dutifully point out to the legislators, if necessary, that this is a most judicious use of tax dollars.

PART III

FORMS OF COMMUNITY-WORKER OWNERSHIP

In addition to ALMCs, worker ownership has recently emerged as a strategy for salvaging threatened companies facing closure. The mechanisms by which employees come to obtain ownership of a business include the following: (1) an offering to employees of the stock of his or her company by a retiring entrepreneur; (2) stock benefits to employees as part of a corporate rewards program designed to motivate or at least satisfy employees; (3) worker buyout of a company as the community faces the crisis of a plant closing. The chapters in Part III focus on the third approach to worker ownership because it is the most relevant to communities in economic distress.

In recent years, there has been a growing trend toward employee ownership in the United States. Giants such as Chrysler are now partially employee owned. Five major airlines and six trucking firms have turned to worker ownership in the past year alone. Weirton Steel, the eighth largest steel producer in America, is 100 percent worker owned, as is the largest sugar producer, U.S. Sugar. Workers in Philadelphia whose jobs disappeared when A&P closed twenty supermarkets in the city are now buying the closed stores and running them as O&O (worker owned and operated) co-ops.

William Foote Whyte and Joseph Blasi discuss trends in the national economy that imply increasing concentration of wealth, less efficient corporate practices, and various myths about big business. They then synthesize a number of cases of worker ownership to provide a flavor of the possible benefits of employee ownership for achieving regional economic stability and job retention.

In Chapter 11, Woodworth and Meek take an in-depth look at the paradoxical treatment of worker ownership in the media. They emphasize the importance of workers not only obtaining ownership on paper, but the need to gain a degree of formal participation in the managing of the workers' firm. The case of the Rath Packing Company in Iowa is analyzed in considerable depth as the authors draw upon several years of action research at Rath. Key turning points are emphasized in the meatcutters' attempt to take control of their company and create a truly democratic organization in the face of monumental problems. Substantive changes in decision making and business outputs are reported, as are the difficult hurdles that make Rath's long-term future questionable.

Finally, in Chapter 12 Woodworth writes of his experience in salvaging jobs through a worker buyout of a General Motors plant in New Jersey. Hyatt Clark Industries is an important development in the employee-ownership movement because of its size, its centrality in a major industry (automobile), its financial success, and its implications for the ability of labor to bargain for more corporate power. The author portrays key issues in the conversion of the plant to worker ownership and the shaping of a new, more democratized work culture. Implications from this buyout are articulated in an effort to assist other groups that may embark on such a venture.

10

THE POTENTIAL OF EMPLOYEE OWNERSHIP

William Foote Whyte
Joseph Blasi

For generations, public policy has been influenced by the ideology of free enterprise: The central notion is that individuals, free to seek their own economic interest in competition with others, would take actions that would contribute to the general welfare of society.

Although no one now visualizes the U.S. economy as responding to the independent actions of thousands or millions of individuals competing in the marketplace, there has been a tendency to personify the business firm and to assume that the firm decides and acts in terms of its own economic interests in a competitive market, just as the individual is assumed to do in the "economic man" model underlying the ideology of free enterprise.

To recognize the falsity of this vision, we need only to review the rapidly growing concentration of economic power.

According to Nadel (1976), "Most of the country's manufacturing assets are in the hands of the 200 largest corporations that held 60 percent in 1972—a jump from 47.7 percent in 1950." He goes on to report, "The third wave of mergers started in 1955 and peaked in 1969. In 1967, 1968, 1969 the merger movement reached all-time high peaks; an average of over 3,500 mergers occurred *annually*. . . . by 1968, 89 percent of all mergers were conglomerate in nature." Nor were the companies that were taken over relatively small. "Between 1962 and 1968, 110 of *Fortune's* largest industrial corporations disappeared as independent companies."

Some spokespersons for business recognize that unsound mergers did take place in "the go-go years of the 1960s" but would have us believe that the merger fever has since subsided. Although the number of mergers has indeed declined, the mergers of the late 1970s are bigger— and not necessarily better—than ever before. In 1975, U.S. companies spent $15 billion to buy other companies. By 1980 that figure had jumped to $40 billion, and in just the first six months of 1981, the amount spent on acquisitions had reached $37.7 billion. If this rate continues, we can say that the money spent on mergers has increased about 500 percent since 1975.

Conglomerate mergers are currently the hottest game in town in the business world. As the merger game continues, small businesspersons, hard-pressed to raise capital at high interest rates, are bound to wonder whether the buying and selling of companies is the most productive way to use our nation's capital.

The most insidious aspect of the conglomerate merger game is that many excutives are forced to play it when they would rather devote their attention to more productive activity. Unfortunately, the most effective strategies for avoiding an unfriendly takeover involve either seeking a friendly merger partner or going out and taking over another company. In either case, defending against a merger results in a merger.

Consider the case of Joseph Gaziano, who told *Fortune* (March 12, 1979), "My dream is simple—to build an empire." As chief executive officer of Tyco Laboratories, a small New Hampshire conglomerate, he made money for himself and his company by buying and selling the stock of other companies. In one series of deals beginning in 1976, Gaziano began buying the stock of Leeds and Northrup. To escape Gaziano, Leeds and Northrup sold a controlling interest to Cutler-Hammer, a company more than three times Tyco's size. Gaziano countered by buying heavily into Cutler-Hammer stock. Cutler-Hammer then arranged to be taken over by Eaton Corporation.

Unsuccessful in these and other takeover efforts, Gaziano forced the target companies to pay him ransom for their freedom in the form of a substantial profit in buying back their stock. *Fortune* estimated that these and other unsuccessful raids yielded Tyco a net profit of well over

$19 million. Gaziano took his failures philosophically, commenting, "Coming in second best isn't so bad if you make a few million dollars each time."

The bonus received by Gaziano was based upon return on stockholders' equity. Although the company's record of operating profits was mediocre and had been declining, the conglomerate merger game enabled Tyco to reach a spectacular 31.8 percent on stockholders' equity for the fiscal year ending May 1978. For this performance, Tyco paid Gaziano $442,514, about half of it as a bonus—not bad for a miniconglomerate.

The conglomerate game has also increased the power and wealth of investment banks and law firms, which make millions of dollars contriving friendly or shotgun marriages. In his important study, economist David Kotz (1978) found that in 1969, 69 of the 200 largest nonfinancial corporations (34. 5 percent) were controlled by banks or other financial institutions. He assumed such control existed if a bank held 5 percent or more of the company's voting stock, or if the bank was the leading supplier of debt capital to a company that relied on debt capital to a substantial extent *and* the financial institution was strongly represented on the company's board of directors.

THE MYTHS OF BIG BUSINESS

Our studies have indicated the falsity of three commonly held beliefs about big business.

(1) Expansion of big corporations creates jobs and is therefore an important factor in reducing the national level of unemployment. At the beginning of 1969, the 1,000 largest corporations provided 20 percent of national employment. In the following eight years through 1976, total national employment rose by more than 9.5 million. What percentage of that increase was contributed by these 1,000 largest corporations?

We asked three classes of Cornell students to guess the correct answer to that question. Their guesses have ranged from 3 percent to 70 percent, with most answers falling between 20 percent and 35 percent. The correct answer? Eight-tenths of 1 percent (Breckinridge, 1978: H1828-1830)!

To be sure, it may be argued that most of the 1,000 largest corporations are engaged in manufacturing, a field that has been providing a decreasing percentage of national employment in recent decades. However, that observation simply helps to explain the figures; it does not change their basic meaning. Whatever else they may do for the U.S. economy, we cannot expect the largest corporations to be a major factor in hiring enough people to produce a substantial decline in unemployment.

At the other extreme, according to David Birch (1979), the smallest firms provide the greatest growth in employment. His figures for the 1969-1976 period show that 52 percent of all jobs added to employment were contributed by independent firms with twenty or fewer employees.

An additional 14 percent were contributed by small businesses holding franchises from larger companies. These days the biggest firms generally grow in number of people they employ only by taking over other firms. In general, do conglomerate takeovers lead to a decline in employment?

It is easy to find large and spectacular cases supporting this thesis. For example, consider the Okonite Company in New Jersey, a manufacturer of cables, which had been a steady profit maker for the 75 years of its independent existence. But then, as a high official of Okonite explained to me, "During the last ten years, we were owned twice by Jimmy Ling. That says it all!" Even under this conglomerate management, Okonite continued to make profits. Nevertheless, when the Ling financial empire collapsed, the 1,800 workers in Okonite would have lost their jobs if the leaders of the subsidiary had not been able to extricate Okonite from the wreckage and reestablish it as an independent (employee-owned) company, with substantial loans from EDA and banks.

In 1969, with the aid of $175 million in bank loans, Lykes Corporation took over a company with capital assets five times as large as its own. In the following years, Lykes reinvested in Youngstown Sheet and Tube Company only a little more than half of what it claimed in depreciation on that subsidiary, using the money saved during good years for steel to buy other companies. When the downturn came, Lykes found itself with an obsolete steel company on its hands, and Lykes' chief executive officer estimated that it would take bank loans of over $550 million in a ten-year period to make the steel company economically viable once more. Because bankers were not inclined to put up that money, Lykes closed the mills, putting 5,000 people out of work in the Youngstown area.

Even when a rapidly expanding conglomerate survives an economic disaster and makes a comeback, we must ask what effects its roller coaster performance has had on employment. Consider the case of the

Whittaker Corporation, as reported by Dan Dorfman (Washington Post, December 16, 1979: F14).

> It started out as an aerospace company in 1947 but then went on to acquire more than a hundred different firms between 1967 and 1970. Reflecting its meteoric growth, Whittaker shares went through the roof rising from 6-3/8 in 1967 to 45-7/8 just a year later. But then the roof caved in (as did its stock) when the company virtually lost control over the management of its various businesses in 1970. It was also the year in which the debt-ridden and highly leveraged company was stung with $8.5 million loss on some $800 million in sales. Whittaker seemed headed for the scrap heap. But then came a major restructuring of the company under a new chief executive (Joe Alibrandi, formerly a senior vice president of Raytheon), the subsequent divestiture of 92 businesses and deeper involvement in the fast-growing hospital management or life science field.

According to Dorfman, this is a story with a happy ending. He adds:

> The stock of the $1 billion company recently surged past 18 in accelerated trading activity. . . . as things stand now, the company's earnings are booming.

Let us now ask questions that are relevant only if one looks beyond stockholder interests. What was the level of employment in these 92 abandoned companies when taken over by Whittaker? What was their level of employment when they were dumped by Whittaker? (Given that the conglomerate had gotten itself into deep trouble before Alibrandi began the divestiture process, we assume that comparison of those two figures would show a substantial loss of jobs during the period of Whittaker ownership of these 92 companies.) How many of these 92 companies went out of business when they were dropped by Whittaker? How many jobs were lost in this going-out-of-business process?

Of course, it is easy to make a case against conglomerate mergers by picking disaster cases, but these negative cases should at least suggest reasons for concern until we have research measuring the employment impacts of mergers nationwide over a period of years. Lacking such research, the best we can do is look to case studies of researchers who were seeking to examine apparently favorable effects as well as unfavorable effects. Beldon H. Daniels and his Harvard associates (1979) made such a study, in which they examined three cases of successful outcomes and five cases of failures. With this selection of cases, of course, one cannot generalize regarding mergers, but, nevertheless, it is interesting to contrast their successes with their failures. Three of the failure cases were outright disasters, with the acquiring firm eventually shutting down its subsidiary (Computer Control Company taken over by Honeywell, Inc.; Youngstown Sheet and Tube taken over by Lykes Corporation; and the Colonial Press, Inc., taken over by Sheller Globe). In the other two failure cases, management people bought back American Safety Razor Company from Philip Morris and reestablished a successful business, and the employees of what became Bates Fabrics saved their jobs with an ESOP buyout. The Daniels group found the three success cases had these conditions in common: (1) Managers of the acquired company had been running a successful and expanding business and willingly turned to a conglomerate as the only way to finance further expansion, and (2) the acquiring company provided the needed capital while allowing its subsidiary to operate in a highly autonomous fashion.

In other words, in the success cases (Gould taken over by Modicon, Secor taken over by Sperry-Rand, and Kendall taken over by Colgate-Palmolive) the researchers found no evidence of the infusion of needed managerial talent into the subsidiary. The conglomerate served a useful function primarily in supplying capital. However, this finding simply underlines the conclusions of other studies indicating that small, profitable, and growing companies have much more difficulty in acquiring capital than large companies. Should we depend upon conglomerates to meet these needs in inefficiently functioning capital markets? Or should we face the problem directly and seek to restructure capital markets?

Beyond their global effects on employment, there is ample evidence of the disruptive effects of conglomerate mergers on local economies. In their unsuccessful efforts to protect Carrier Corporation from a takeover by United Technologies, business and community leaders expressed alarm over the potential impact of the merger on the economy of Syracuse:

> The Chairman of Syracuse's Small Business Council estimated that 70-90 small businesses might close if Carrier changed to centralized purchasing, since 42 Syracuse companies were 90% dependent upon Carrier, and another 40 were 50% dependent [Daniels et al., 1979: 352].

That such concern was well founded is illustrated by what happened to acquired firms in other states. Citing a 1971 study by S.L. Brue, the Daniels group reports that in Nebraska 45 percent of the acquired firms switched to centralized purchasing, 75 percent abandoned some or all of the services of local financial institutions, and 77 percent used the legal services of the parent corporation. Similarly, the Daniels group reports a 1969 study by John D. Udell indicating that for the state of Wisconsin 75 percent of acquired firms changed legal services to the parent company, 70 percent of acquired firms changed to financial institutions of the parent company, and 70 percent of acquired firms changed accounting firms.

(2) A big company would not shut down a plant that was making money. Of course, many of the plants shut down have been losing money, yet we have found cases of consistent profit-makers that have nevertheless been closed. How can that be?

A plant producing library furniture in Herkimer, New York, had made a profit nineteen of the twenty years it had been owned by Sperry-Rand, yet it had not been yielding the approximately 20 percent that the conglomerate decision makers expect on their invested capital, and, furthermore, it did not fit into the business strategy of the chief executive officer. Sperry-Rand officials estimated they could get more for their money by shutting down the plant, declaring a tax loss, and selling the machinery than they could by continuing to own and operate that plant.

For generations the name of Bates had been a symbol of top-quality fabrics, yet in 1977 management decided to shut down its sole remaining textile mill in Lewiston, Maine. In the 1960s, the Bates Company had begun to acquire companies in the fields of energy and natural resources. Finding that these firms yielded around 15 percent on invested capital, whereas Bates could expect only 5 percent to 7 percent in textiles, the conglomerate leaders made the rational decision to close the Lewiston plant and shift their capital to more lucrative fields.

What was rational to the conglomerate made no sense to the 1,100 workers of the Lewiston plant or to local business and community leaders, faced with loss of the largest payroll and the largest source of taxes in their city. With the aid of a loan guarantee from the Farmers' Home Administration, the employees and local management people secured bank credit (at 10¼ percent) to buy the plant and save the jobs. At the time, management spokespersons estimated that they would be able to make substantial investments in plant modernization, meet the loan payments, and make a profit of 5 percent to 7 percent.

Incidentally, this case provides an answer to a question often asked us: If the plant could be expected to make a profit, why didn't some private investors come along and buy it? If he or she had no interest in jobs and business in Lewiston, what private investor would consider making an investment involving a substantial risk and promising to yield 5 percent to 7 percent return? In contrast, from the standpoint of the workers and their fellow citizens of Lewiston, a return of 5 percent to 7 percent on an investment that saves 1,100 jobs and maintains a major source of local taxes and local business is an attractive proposition indeed.

Finally, consider again the Okonite Company, which had been a steady profit-maker. When the conglomerate that had taken it over ten years earlier went bankrupt, top management would have had no choice but to shut Okonite down—if workers, local management, and company officials had not worked together to extricate Okonite through transformation of ownership.

(3) Big business is efficient. Therefore, you cannot expect local workers and managers to make a go of an operation the performance of which did not satisfy the decision makers of the big company. Fortunately, we need spend little time in debunking this myth since the Japanese ability to outperform companies in some of our major industries has shaken the confidence of U.S. business leaders themselves and has left few believers among the general public. In the years immediately following World War II, the wartime performance of our "great arsenal of democracy" excited worldwide admiration, and productivity teams came to us from Europe and Japan seeking the secret of U.S. management know-how. Now U.S. business leaders are going abroad, seeking foreign know-how.

Two Harvard Business School professors, Robert Hayes and William Abernathy, precipitated an expanding current of revisionist thinking regarding American management (Hayes and Abernathy, 1980). Their article evoked an extraordinary response in the business and academic communities, bringing in more purchases of reprints than any other article published in the history of *Harvard Business Review.* The authors have since become much sought-after consultants and speakers at management conferences.

The Hayes-Abernathy critique emphasizes the following points:

(1) Top management people have become so fixated upon financial analysis geared to short-term (quarterly) "bottom-line" results as to neglect the needs for progress in research and development and production.

(2) This concentration on financial analysis and manipulation has led to an increasing tendency for big companies to appoint lawyers and finance specialists to their chief executive officer positions. This naturally reinforces neglect of the operational problems that must be solved if the company is to hold its own in international competition.

CAN WORKER-OWNED FIRMS SUCCEED?

Since the record is now clear, we no longer need to answer the question: Can conversion to employee or employee-community ownership save jobs? Still, when we tell our story, many people find it very mysterious that such a firm can succeed where big business has abandoned the plant. Although explanations for success will vary from case to case, we are already finding important common elements.

(1) Escape from remote-control decisions. In the case of the precipitous drop in volume of business and profits following one conglomerate takeover, we have documented a series of bad decisions. One example should serve as an illustration. Cluett-Peabody, after acquiring Van Raalte, decided to save money in marketing by dropping Van Raalte salespeople and having their own salespeople add the Van Raalte line to their wares. Although this saved selling expenses, it had a drastic impact on Van Raalte sales. The station wagons the salespeople used were not large enough to contain samples of all the Cluett-Peabody lines plus the Van Raalte products, so some of the Van Raalte items were left behind. Furthermore, Cluett-Peabody salespeople did not know the Van Raalte merchandise as they did the lines in which they had much experience. Thus the whole selling campaign broke down.

When Saratoga Knitting Mill, which was originally part of the Van Raalte Complex, became employee owned and independent from Cluett-Peaboby, President Donald Cox contracted with a marketing firm that mounted a successful campaign to sell the fabrics produced at Saratoga Springs in the markets for which they were especially designed.

(2) Restraints imposed by headquarters. Consider the case of the Library Bureau plant that belonged to Sperry-Rand. The plant had always had its own sales force and was not dependent upon Sperry-Rand for its market. In fact, being part of the conglomerate imposed serious barriers in marketing. For example, it was a rule of Sperry-Rand that the Library Bureau salespeople could not call on any customers served by Sperry-Rand. Although this left to the Library Bureau its main markets with public libraries and those of educational institutions, the rule barred the plant from selling to a large number of industrial and business firms that used library equipment. The Library Bureau could only enter these markets through Sperry-Rand salespeople, who were unfamiliar with Library Bureau products and had more important things to sell. The handicaps were similar in the export field. The

Library Bureau could enter the export market only through the international division of Sperry-Rand. Robert May, formerly head of sales for the Library Bureau and later president of Mohawk Valley Community Corporation, explains what would happen in these words:

> We were not officially barred from exporting, but to sell anything outside of the country, we had to send our proposal to the international division, and it would just die there. We would never hear anything back [personal communication].

(3) "Modernizing" the job shop. Conglomerate executives and staff specialists are likely to be graduates of business schools, where they have learned from books designed for mass production operations and, therefore, apparently requiring standardized systems of organization and control. Thus it is often the case that when a conglomerate acquires a job shop operation that produces a wide variety of items, one-by-one or in small batches, that corporate executives are appalled by the apparent confusion and disorganization. Furthermore, they are likely to have had their chief experience in operations yielding a much higher margin of profit than is produced by the job shop. Such individuals therefore have a natural tendency to assume that the profit differential is due to the superior management skills of their own people rather than being due to the inherent nature of different types of industrial operations or to the nature of the market. They therefore assume that if they "modernize" the job shop, imposing the systems of organization and control that they have used in mass production operations, they will be able to turn the plant into a really efficient unit (see Meek, 1983: 60-63).

The results of acting on this mystique of the management generalist are often disastrous. For example, when the now employee-owned firm, Jamestown Metal Products, was owned by the conglomerate AVM (American Voting Machines), top management sought to plan and control production scheduling and work flow within the plant through the use of a computer. That operation required two employees within the plant and one at the home office, plus substantial costs in computer time. Local management complained that the computer gave them nothing but garbage, but top management rejected their "old-fashioned ideas." Now that the plant is employee owned, the records necessary for production scheduling and work flow are maintained by a single employee who works half time on other assignments. And the now unmodernized company is doing very well, paying out substantial amounts in profit sharing to workers and currently having a book value many times what the figure was at the time of the transfer of ownership in 1973.

We recognize that the computer can be useful and even invaluable in some types of business operations, but its efficiency depends upon the cost of maintaining and producing the information, as well as the potential for routinizing the activities it monitors or controls.

Consider this hypothetical case that is representative of many job shops. The plant has seven different types of machines, with varying numbers of each type. Even within machines of the same type, there is variation in age, speed of operation, and quality of performance. The plant produces a large number of *different* items in single units or in small batches. Some items are worked on by several types of machines, and some items require the unit to go from machine type A through B and C and then back to A, and so on. Performance depends upon the skill and diligence of workers and upon scheduling, so as to keep machines and workers busy, reducing machine down time or worker waiting time, and so on.

In this situation, the number of different paths a given unit might follow from machine to machine is well nigh infinite. This is not to say that it would be impossible to develop a good computer program to schedule production and work flow. However, the complexity of the task is so enormous as to require intensive and extended studies in industrial engineering. Even then, whenever a change in the market occurs, the old programs would become obsolete, and new studies would be required. Therefore, although it is possible to develop good computer programs for such operations, the costs far outweigh the benefits.

Without computers, how do job shop people make decisions on these scheduling and work-flow problems? Their decisions are partly intuitive and partly carefully reasoned but, in any case, are based upon past experience and intimate familiarity with the characteristics of workers and machines.

This is not to say that management skill and experience are unimportant for an employee-owned company. We are simply arguing against the mystique of the management generalist: the idea that a good executive in one situation will automatically be a good executive in a quite different situation. The effective executive in the employee-owned firm will be someone who, in addition to what he or she has learned in formal management courses, has a wealth of experience in the particular line of production and marketing in which the firm is engaged.

(4) Shedding corporate overhead. When the Library Bureau was part of Sperry-Rand, the Herkimer plant was contributing $600,000 a year in

corporate overhead. When we asked John Ladd, a member of the board of the Mohawk Valley Corporation, and, as director of Mohawk Valley Economic Development District, the key figure in converting ownership of the plant, how much value he thought the Herkimer plant had been receiving in return for that $600,000, he replied, "I would say it was worth about $1.50."

Although that figure may have been somewhat of an underestimate, it does point to a general problem. When a branch plant is one among a number of plants producing similar items and having similar problems, corporate management can provide many services to those branch plants. When the plant in question is one of a kind or fits into a very small part of a conglomerate the major interests of which are in other lines, then the services provided by corporate headquarters are unlikely to be worth the cost. Thus the new company improves its profit position substantially by getting out from under the corporate overhead burden.

(5) Improvements in employee motivation and productivity. In a study commissioned by the Economic Development Administration, Michael Conte and Arnold Tannenbaum have found, in general, that employee-owned firms tend to show superior records both in job satisfaction and in financial performance. In thirty firms from which they obtained profit figures, Conte and Tannenbaum report

> that employee owned firms can function efficiently and profitably. Furthermore, analyses concerning the possible determinants of profitability of these 30 companies indicate that the single most important correlate of profitablility among the aspects of ownership that we measured is the percent of the company's equity owned by non-managerial employees. The greater this percent, the greater the profitability of the firm [1977: 2-3].

(6) Reducing the costs of controlling worker behavior. Although employee motivation seems an intangible item, as we get down to cases, we can link it with concrete economic results. Consider these examples from Saratoga Knitting Mill. When under outside ownership, the plant employed three janitors. Now the employees keep their own work areas clear and neat, there is only a part-time janitor, and the factory floor is as clean as it has ever been. Prior to the ownership change, the pilferage problem was serious enough to require a guard service. This service was terminated following the purchase, yet the incidence of pilferage has been reduced to almost zero.

Another imporant gain was made in the reduction of fabric loss. Apart from pilferage, fabric is lost through the trimming of edges, damage, and so on. Under conglomerate ownership, a 1 percent allocation of the total manufacturing cost was budgeted for fabric loss. During the first year of operation only .25 percent of SKM's net sales was incurred as a loss of fabric. Similarly, in the textile industry, an average of 20 percent of batches have to be redyed because the color is not the correct shade or there are inconsistencies in color. Consequently, the conglomerate management allowed for this with a budget allocation of 3 percent of manufacturing costs. With increased quality concern among the employees the redye costs for the first year of SKM were only .25 percent of net sales. In the words of one foreman, "Now that they own the plant, they take care of getting the colors closer."

There is also a large potential for savings in the costs of supervision. In the traditional private firm the foreman is expected not only to be a coordinator and work scheduler but also a policeman, seeing to it that workers obey the rules, meet company standards on output and quality, keep working diligently, and so on. To the extent that the policing functions can be dispensed with, the firm needs fewer supervisors. This is shown strikingly as we compare supervision in the plywood worker cooperatives and in comparable private firms. Greenberg (1981) reports one case in which a private company bought out a cooperative and immediately hired seven additional supervisors. He estimates that, in general, private plywood firms have four times as many supervisors as worker cooperatives of the same size. Nor is the reduction in supervisory costs achieved at the expense of worker production. As Paul Bernstein and others have reported, the plywood cooperatives in recent years have far exceeded the productivity performance of private plywood plants.

(7) Trying harder to survive. Compared with top management of conglomerates, executives of employee-owned companies will not be so likely to milk the profits from one plant in order to shift capital to more profitable plants or subsidiaries. In the first place, the employee-owned company is much less likely to have a number of subsidiaries, which makes it possible to shift profits from one unit to another. Furthermore, the employee owners are in a much better position than workers in a privately owned corporation to protest any management disinvestment decision that would threaten their jobs.

Although we cannot assume that the employee-owned firm would always be successful in meeting foreign competition, workers will naturally refuse to export their capital and their jobs to other countries. In the face of competition, some employee-owned firms may go bankrupt, but closing of plants will not be due simply to a management determination that more profits can be made by shifting production to Mexico, Singapore, or South Korea. In other words, the employee owners will try harder to meet the competition because their alternative is unemployment.

CONCLUSION

As can be seen from the foregoing discussion, the smaller locally controlled and employee-owned firm can in a variety of cases be more economically successful and certainly more socially responsible in managing its resources than the large firm. Where thousands of jobs are currently being lost in the United States due to mismanagement and corporate divestiture, many viable firms could in fact be saved, and not only maintain current jobs, but even produce new employment through employee ownership. This innovative and potentially important development needs to be incorporated into a national industrial strategy for strengthening and revitalizing U.S. industry.

REFERENCES

BIRCH, D. (1979) The Job Generation Process. Cambridge: MIT Program on Neighborhood and Regional Change.

BRECKINRIDGE, J. (1978) "Report on a study carried out for the House Sub-Committee on Small Business, Anti-Trust, Consumers, and Employment." Congressional Record (March 8).

CONTE, M. and A. TANNENBAUM (1977) "Employee ownership: report to the Economic Development Administration" (project 99-6-09433). Washington, DC: U.S. Department of Commerce.

DANIELS, B., K. BOBIOW, B. SIEGEL, and A. VERRILLI (1979) "Preliminary exploration of the impact of acquisition on the acquired firm" (discussion craft). Cambridge, MA: Harvard University, Department of City and Regional Planning.

GREENBERG, E. (1981) "Industrial self-management and political attitudes." American Political Science Review 75 (March).

HAYES, R. and W. ABERNATHY (1980) "Making our way to economic decline." Harvard Business Review (July-August).

KOTZ, D. (1978) Bank Control of Large Corporations in the United States. Berkeley: Univ. of California Press.

MEEK, C. (1983) "Labor management cooperation and economic revitalization: the story of the growth and development of the Jamestown Area Labor-Management committee." Ph.D. thesis, Cornell University, State School of Industrial and Labor Relations.

NADEL, M. (1976) Corporations and Public Accountability. Lexington, MA: D. C. Heath.

11

WORKER-COMMUNITY COLLABORATION
AND OWNERSHIP

Christopher Meek
Warner Woodworth

Employee ownership appears to be an idea favored by many American
workers. A 1975 national survey conducted by Hart Research found
that 66 percent of those sampled indicated that they would prefer
working in an employee-owned firm over a company owned by either
private investors or the government. Yet few employees have ever
actually had the opportunity to work for a firm in which they have been
part-owners. The number, however, has grown considerably over the
past ten years, and the trend seems likely to continue throughout the
1980s. Political leaders ranging from Orrin Hatch and Ronald Reagan
on the right to Senator Edward Kennedy on the left have been greatly
impressed by the trend and have given their backing to legislative
support of this movement at the federal level. Senator Russell Long of
Louisiana has been a leading advocate of employee stock ownership
plans (ESOPs) as a means of solving the productivity problems of the
United States and has conducted a study of ESOP firms. He reports that

the 22 companies that have responded to my request for information have advised me that since establishing their ESOP, their average pretax profits have increased 125 percent; their sales volume is up 67 percent; their sales per employee has increased 38 percent; and the number of employees employed by these companies has increased by 30 percent [House Small Business Committee, 1979: 176].

THE ROOTS OF DISSENSION

Unfortunately, success stories, when examined over time, have in some cases proven to contain serious built-in flaws from the standpoint of industrial relations—flaws that over the long run can make a firm saved through employee ownership a virtual union-management "time bomb" (Blasi, 1982: 6). This has especially been the case with ESOPs and the reason is their failure to vest employees with the full rights of ownership.

A 1980 General Accounting Office (GAO) study of thirteen ESOPs in the employee-owned companies found that the stock plans were not being operated in the best interests of their employee-owner participants. It was discovered that the majority of the firms were using inaccurate stock valuation procedures and had failed to provide any stock repurchase provisions. Furthermore, the GAO found that all but one of the companies had, through various means, succeeded in establishing their plans without giving the participants full voting rights. One form simply issued employees nonvoting stock. Most of the other twelve placed an employer-selected and -appointed committee in charge of actually voting the employees' stock. Only one of the twelve required that the employer-appointed committee vote the plan's stock in strict accordance with the instruction of plan participants (GAO, 1980).

OWNERS IN NAME ONLY?

It is this failure to recognize their employee participants as true owners and afford them the full rights and privileges of shareholdership that makes ESOP firms potential labor-relations time bombs. Generally, when conversion to employee ownership occurs, union leaders and rank-and-file employees are euphoric over the saving of their jobs and give little thought to their rights as owners. Research by Cornell's New Systems of Work and Participation Program has suggested that the good feelings stemming from such euphoria represent the first of two stages in the development of trade union-management relations in the typical firm saved through employee ownership (Whyte, 1978).

Because of their failure to confront and clarify their respective assumptions about what employee ownership should mean in terms of industrial relations, both parties unwittingly set the stage for the second phase of union-management relations, which is marked by rising distrust and the reemergence of old frictions and conflicts, as well as several new ones. As the relationship gradually deteriorates, so does productivity. Workers begin to realize they are essentially owners in name only, with no more influence on either daily or major corporate decisions than they had prior to the conversion to employee ownership. Shifts from euphoric enthusiasm to distrust and hostility are well illustrated by the changing attitude of Carl Vogel, union president for the factory work force at the Mohawk Valley Community Corporation—a firm saved from closure under Sperry-Rand in the fall of 1976 through employee and community ownership. Shortly after the conversion, Vogel spoke with great enthusiasm of the dawning of a new union-management relationship. But within only fifteen months he was saying, "You want to know what working in that plant means to the workers now? Ask anybody. They'll tell you the same thing: 'I've got a job. Nothing else.' Labor relations are no better than they were" (Whyte, 1978: 80).

As time passed, Carl Vogel's anger intensified and his views about the lack of positive change in industrial relations crystallized beyond bitterness and frustration to a recognition of the need for employee involvement in company decision making and problem solving. Several years after the conversion to employee ownership he made this point in an interview with *Washington Post* journalist Daniel Zwerdling:

> But now, as he sits in his living room one rainy evening four years later, union president Carl Vogel is talking angrily about how the dream of worker ownership soured. "We saved our jobs yes, but it could have been a lot better," Vogel says, clenching his fist for emphasis, "but they (management) just kind of pushed us to the side and said, 'We're the boss and we're going to run the show.'
>
> "The people on the floor know a lot more than running machines," Vogel says, thumping the couch. "We know how to make this company run."

That Vogel developed such feelings is not surprising given the fact that after its first year of successful performance, the firm began to slide.

By the close of its second year as an employee-owned company, it had
lost $1,868,000 due in large part to extremely low bidding on sub-
contract work (Ross, 1980: 111). Furthermore, the firm's management
had consistently ignored the suggestions of union employees for
improving the efficiency of their old, turn-of-the-century multistory
factory. The need to cut costs has resulted in a reduction of the work
force from 250 persons to only 100 workers today. And the union's
feelings of resentment toward management heightened recently after the
board issued enough new shares of stock at 50 cents per share (they sold
for $2.05 per share during the takeover in 1976) to enable four members
of management to obtain 50 percent of the corporation's voting stock
and thus hold total control.

OWNERS ON STRIKE

Similar industrial relations problems have also arisen at a number of
other employee-owned firms, accompanied in some instances by
considerably more dramatic forms of strife than occurred at the Mohawk Valley Community Corporation. At the Okonite Corporation of
Ramsey, New Jersey, a 100 percent employee-owned company in which
employees have no voting rights, a 12-week strike took place during the
spring of 1981.

During the summer of 1981, a strike also occurred at the South Bend
Lathe Company, which has frequently been cited by ESOP advocates as
an economic success story. And indeed, over the five years that followed
the ESOP creation in 1975, net annual sales nearly tripled, and
production employees received wage increases every year as well as four
production bonuses. Workers nonetheless became frustrated and
dissatisfied with their lack of influence in daily operations and overall
policy. Richard Boulis, the company president—who personally appointed all the trustees voting the ESOP stock—had essentially continued to run the firm as he had before the conversion to employee
ownership. It was not until four years after the transition to employee
ownership and many complaints about the lack of communication with
workers that Boulis finally agreed to hold a monthly information
meeting with 35 members of the rank and file. In this meeting he
discusses the state of the business except for those matters he considers
"strictly confidential" (Ross, 1980: 110).

The first signs of employee frustration emerged in 1977, when the
work force voted to reaffiliate with the United Steelworkers, from

whom they had broken at the inception of the ESOP. In August 1980, the workers went on strike for nine weeks, demanding cost-of-living increases and carrying signs on the picket line reading "ESOP's Fables" and "Owners on Strike." The feelings behind these signs were particularly well articulated in a statement made to *Washington Post* reporter Warren Brown during the strike. Randy Reynolds, a machinist who had played a key role in convincing workers to accept the ESOP when the idea was first proposed for saving the company, explained:

> What we have had there for the last five years is ownership without control. . . . We've bent over backwards since 1975 to make a good product and keep it selling. . . . We've kept our mouths shut—covered up our differences with management to avoid publicity. . . . But all we got was the same treatment we had before the ESOP, maybe even worse. We make no decisions. We have no voice. We're owners in name only.

In response to the leveling of such criticisms, Richard Boulis told Brown:

> Employee ownership does not mean employee management. Somebody has to give orders to make things happen. You can't run a business by committee. . . .

> When we bought this company, I didn't have time for human relations. I didn't have time to go around patting people on their butts. . . . I didn't think about anything except keeping this business going and making money. But now, maybe I'll find some time for human relations. I guess I'll have to [Washington Post, September 30, 1980].

Considering the tone of Mr. Boulis's comments, it seems doubtful that any substantial increase in employee participation will take place in the near future. Likewise, recent developments also seem to rule out the possibility of institutionalizing some degree of employee-control altogether. In anticipation of 100 percent employee ownership, resulting from full vesting of the ESOP over the next couple of years, the firm's president used company funds to purchase stock.

RATH PACKING COMPANY: TRYING TO FORGE ANOTHER WAY

In contrast to the ESOP conversions that predated it, the Rath Packing Company's switch to employee ownership was spearheaded by the local union leadership, with the support of their rank-and-file members. They were the principal creators and driving force behind the structuring of the Rath ESOP and the development of employee participation from the shop floor to the board of directors. It is this key difference in leadership that strikingly distinguishes Rath from earlier cases and seems to have made the crucial difference of employee rights.

COMPANY BACKGROUND

The offices of the Rath Packing Company's headquarters and major slaughtering and processing operations are located in Waterloo, Iowa, in the northeastern corner of the state. The greater metropolitan area, which the city of Waterloo shares with the adjoining community of Cedar Falls, Iowa, has a population of approximately 1900 persons working in its slaughtering and processing operations and 200 office staff at its corporate headquarters.

One of Iowa's oldest packing companies, Rath was first established in the Waterloo area in 1891 by two cousins, J. W. Rath and E. F. Rath, with a modest capital investment of $25,000. It had only 22 employees at that time.

By the 1940s, Rath's Waterloo facilities had become the world's largest and most modern packing house. On a 37-acre site, the Waterloo plant was by then a 7-story structure with approximately 2 million feet of floor space. Its slaughtering operation alone killed roughly 12,000 animals per day, including hogs, beef cattle, and sheep. At its peak, the company employed nearly 8,000 workers. With this growth, the Rath organization began to lose the sense of family that had once existed during its early years. As a result, unionization occurred and union-management disputes became a frequently recurring problem beginning in the late 1940s, when a violence-ridden strike took place, and resulted in the death of a union picket who was shot by a nonunion strike breaker.

During the 1950s, Rath's position as a major packer declined as larger conglomerates moved into the marketplace. With their easy access to quick cash, they drove the company out of the beef and sheep markets. In addition, while other companies aggressively sought the business of the growing supermarket chains during the 1950s, Rath waited until well into the early 1960s to take this path. By that time, its competitors had a significant portion of the market under control.

Such shortsightedness was apparently characteristic of Rath management during that period. Unlike most major packers, Rath did not reinvest significantly in the development of modern plants and equipment. Management feared the cost of severance payments that would accompany automation and the construction of a more efficient single-story plant. By 1972, the decision to not modernize caught up with Rath. When combined with a lower volume of hogs and higher prices, as well as the Nixon administration's price controls, the company was driven into the red with a $14.6 million loss on $277 million in sales. Rath's performance did not improve afterward. In the 10-year period between 1969 and the close of 1978, Rath lost over $20 million, and it was consistently in the red from 1975 through 1978.

THE EMERGENCE OF EMPLOYEE OWNERSHIP

After so many years of heavy losses, Rath was near bankruptcy, and this was not the first time. Because of its unstable financial situation, the company was unable to obtain a line of credit from normal lending institutions in the early 1970s. Rath was therefore forced to turn to an industrial high-risk lender or "factor" for credit who charged a rate of 5 to 6 percent over prime. Of course, borrowing working capital at such high rates of interest imposed a heavy burden, and the precariousness of Rath's situation increased eventually to the point where the lender reduced its line of credit substantially after the firm sustained a loss of $6 million in 1975. In an effort to avert bankruptcy, Rath management appealed to the National Bank of Waterloo for assistance, and the bank arranged a 90 percent loan guarantee through the Economic Development Administration of the U. S. Department of Commerce. As a result, Rath received total loans of $6 million, with $4 million from a Pittsburg bank and the rest from local banks.

By 1978 and after three more years of losses, Rath again faced bankruptcy. In an effort to forestall this event, management this time approached the union for help through the reduction of labor costs. On May 11, 1978, Emmett McGuire, then company president, presented a letter to Lyle Taylor, president of the Waterloo plant's union, Local P-46 of the United Food and Commercial Workers Union (UFCW). The letter proposed that all UFCW employees take a $.50 per hour cut in wages and give up two weeks of their annual paid vacation. Shortly after receiving management's proposal, Taylor presented the demands to his membership at their rank-and-file meeting, and the reaction was overwhelmingly negative because of a lack of faith in management.

Faced by Local P-46's rejection, both management and local government officials concluded that bankruptcy was inevitable, and in an
effort to soften the blow, a community committee was organized composed of Rath management, union officials, and political leaders under
the auspices of the Iowa Northland Regional Council of Governments
(INRCOG). Their role changed rather quickly, however, when at the
group's first meeting Lyle Taylor abruptly announced that he and
Local 46 were "not ready for a funeral for the Rath Packing Company."
He proposed that the newly formed committee work together in a
collaborative effort to find a way to save the company. The group then
decided to once again approach EDA for help with interim working
capital. It was also decided that the City of Waterloo would apply for an
Urban Development Action Grant (UDAG) from the U.S. Department
of Housing and Urban Development (HUD), which they could, in turn,
loan to Rath at a low interest rate for badly needed modernization of the
Waterloo plant.

EDA responded favorably, but refused to issue a loan without a
feasibility study. The agency provided a small grant to finance the study,
and two consulting firms approved by EDA did the analysis. The study
was carried out during the latter half of July and all of August 1978.
During this period, they examined the condition of the Waterloo plant,
evaluated Rath's sales and marketing program and analyzed the
performance of both management and the work force. The consultants
also estimated the impact of a Rath closure upon the surrounding
community and arrived at the following provisions:

(1) Rath's employees, including all of its 2,179 workers in the Waterloo plant
    and offices, would lose their jobs. In Waterloo alone a $36.4 million
    payroll and $15.5 million in fringe benefits would be lost.
(2) Although retirement benefits under the Employment Retirement Se-
    curity Act (ERISA) would be continued, it would likely be at a
    significant reduction.
(3) The unemployment level for Black Hawk County would immediately
    increase to 8.1 percent with the potential of reaching 15 percent due to the
    "ripple effect."
(4) Rath stockholders would lose their equity.
(5) Payments of outstanding debts would be long delayed and would not
    exceed $.50 on the dollar.

In terms of survival, the consultants concluded that Rath could be
revitalized if several short-term actions were taken. Some of the actions
recommended were as follows:

(1) That immediate action be taken to obtain funds to meet a pension
    payment of $5,860,000 due September 15, 1978, in order to avert the
    company being thrown into bankruptcy.

(2) Elimination of inefficient labor practices and operational procedures in the Waterloo plant and at Rath's corporate offices at an annualized savings of $3.4 million per year.

(3) Reductions and reorganization of the sales force and the management staff and improvements in the management information system at a cost of $1.4 million and an annual yield of $3.4 million in savings.

(4) For capital improvements and plant modernization of $4.3 million at the Waterloo facility the consultants estimated yearly savings of $6.4 million could be achieved (Gunn, 1980: 89).

The consultants concluded that these measures would ensure Rath's survival for the short- but not long-term. They explained that the company would have to construct a projected cost of $32 million to $52 million.

Immediately after receiving the consultants' report, work began on implementing the short-term recommendations. In September of 1978, Local P-46 agreed to support an IRS waiver on the $5.8 million pension payment. Two months later, the union membership agreed to eliminate piece-rate incentives.

Management also took action on some of the consultants' recommendations by substantially reducing the sales and administrative staffs in October 1978. The capital improvement recommendations were also integrated into a proposal to be submitted to HUD for obtaining a UDAG grant to the city of Waterloo. The amount requested in the proposal was that computed by the EDA consultants for making their suggested short-term improvements or, for example, $4.3 million.

During this same period, Black Hawk County received a $3 million EDA grant, which it in turn loaned to Rath for interim working-capital through a nonprofit corporation newly formed to administer the funds—the Black Hawk County Economic Development Committee, Inc. (BEDC). The benefit of EDA funds was multiplied through this approach because it created a local revolving loan fund that could be used to make other low-interest loans to enterprises in the Waterloo area as Rath paid back its debt.

After receiving the EDA grant and finishing the UDAG application for capital improvement funds, BEDC then sought a new CEO as recommended by the consulting team. EDA strongly urged Rath to hire another consultant for this position, Jack Sheehan of East Cliff Associates. Local P-46 leaders expressed strong opposition to the hiring of Sheehan once his plan for reviving the company was laid on the table. Their opposition was fueled by the following proposals, which were central to Sheehan's plan:

(1) East Cliff Associates would negotiate a new three-year contract with Rath production employees that would include a 25 percent reduction in

wages and benefits (roughly a $4 per hour cut). Included in the proposal was a profit-sharing plan that would share with Rath employees 40 percent of all pretax profits.

(2) Use of the new union contract to induce Rath's lenders to substantially increase the firm's line of credit.

(3) Upon approval of Rath's stockholders, Mr. Sheehan would be given an option to buy control of the company through purchasing 1.8 million shares of stock. The shares, which were then selling for $4.00 would be sold to Sheehan for $1.00 per share.

The leaders of Local P-46 reacted quickly and negatively to Sheehan's proposal and countered with a surprising plan of their own that put them in direct competition with him. In an unprecedented action for a local union, they proposed that Local 46 bargain for the establishment of an employee-ownership plan. Lyle Taylor explains the development of this turn of events:

> Things got kind of slow and all of a sudden there was another group of consultants that showed up, and everybody that this guy [Sheehan] talked to became convinced that he could turn the company around. . . . One of his associates called me, and I met with him, and they talked generalities. . . . I told them to put it down in writing. . . .

> Well, they came back with some city officials. They had the backing of EDA and some government agencies to come in and do something. They had a good track record of turning companies around, and then when they brought the real white knight [Sheehan] in with all his ideas, and he laid them out, and we listened; we put a pencil to it, and we didn't see any magic in that other than an employee "give-up." . . . It was kind of put on the basis of well, "This is where you will be and that's better than unemployment, and ADC [welfare] . . . "

> We tried to negotiate with these guys to see what there was to bargain with and they would not bargain. Sheehan's first offer was what he was going to do. Then they notified us that they had rented the largest auditorium in the city, and they were going to have a meeting with all our membership, and . . . go right past the union by putting their proposal straight to the people.

> I guess it was at that point Chuck Mueller and I worked to come up with some alternatives. Really the only alternative we could come up with was to take a stand and go for employee ownership. Actually the company had offered to sell our people the company in a letter a little before this when they saw we weren't going to go for their original "give-up" plan.

> The only reason they brought up the idea of trying to buy the company then was because they were trying to be cute with me. I had played those games before with people who want to play games. So I sent them a

committee to talk about buying the place, and, of course, I set up my vice president and coached him on what he was to say and who he was to meet with. But then I got the call, and they said, "No, we want to meet and talk about these other things . . . " When these consultants came in, though, and offered to buy the company, everything happened real fast and so we had the employee ownership to fall back on. So then Chuck and I drafted a purchase proposal to take over the company and buy their stock. . . . We felt with our plan it was us, instead of his plant it was our plant. Of course, the executive committee decided to support what Chuck and I felt was a sound decision.

On March 7, 1979, the union dropped a bombshell when Taylor sent a preliminary purchase offer to Emmett McGuire, president of Rath. Management and local government officials were shocked at what seemed to be a preposterous idea—that the workers provide the equity capital to keep Rath afloat and turn the organization around. Even Taylor and Mueller's international union regarded the idea as bizarre. (They did not interfere, however—their key concern was that Rath workers not break the national bargaining agreement.) But with further discussion the idea began to gather steam.

Events culminated in June 1980, when the workers voted in union elections to buy the company and the public stockholders approved the plan and the sale.

THE PLAN

When the leaders of Local 46 first proposed to buy Rath Packing Company, they had only a vague idea of how this concept should be implemented. However, by June 1979, when a formal agreement was signed, the idea had crystallized into a well-structured plan with the help and advice of three professors from Cornell University—William Foote Whyte, Tove Hammer, and Robert Stern. Unlike the previous efforts, the Local 46 plan placed equal priority upon employee control and saving the company and, as a result, a milestone step was taken toward the guarantee of employee/shareholder rights.

The financial aspects of the Rath plan essentially worked this way:

(1) Rath workers would purchase a block of 1.8 million shares of Rath stock at the price of $2.00 per share through a payroll deduction plan by deferring $20 per week from their paychecks for stock purchase.

(2) Rath workers would take a 50 percent reduction in their paid vacation time and holidays.

(3) Rath sick pay had been paid from the very first day of sickness. Now the first three sick days would not be eligible for sick pay benefits.

(4) A total of $.45 in cost-of-living wage increases were deferred until the company showed sufficient sustained profits.

(5) A union escrow account was established to hold all wage and benefit deferrals, until such time that Rath stockholders had ratified the stock purchase agreement. When the proposal passed, the funds were turned over to the company and the escrow account closed with continued deferral of benefits.

(6) In the event that the corporation made a profit, 50 percent of all pretax profits would be assigned to an employee profit-sharing fund to restore funding to the pension plan and pay employees deferred wages and benefits.

(7) All profits remaining after profit sharing and taxes would be used for corporate capital improvements. No dividends would be paid during the term of the Local agreement.

(8) A union appointed certified public accountant would have access to company financial records as frequently as twice a year.

As can be seen, the above cited measures served to generate substantial new capital resources. However, as important as these measures were in sustaining the company, the key distinguishing factor in the Rath conversion to employee ownership was the nature of the ESOP trust itself. One key feature established to provide worker control was an agreement that the company's board of directors would be enlarged from six to sixteen once the stock purchase process had started. It was agreed that the ten new directors would initially be appointed by the union and then afterward by the ESOP trustees.

Another unique feature of the Rath ownership plan was the internal decision-making structure of the trust. A five-worker board of trustees was established and elected by the ESOP participants on the basis of the pure cooperative principle of one person/one vote. Unlike other ESOPs, the plan was set up so that each participant would have only one vote, and the trustees would be required to vote in the general stockholders' meeting in accordance with the instructions given them by the plan participants through majority vote.

As a group, the five trustees were delegated responsibility for administering the ownership plan and voting the stock of all employee participants at the annual stockholders' meeting. In May 1982, purchase of the total 1.8 million shares was complete and, had all of these shares been committed to the ESOP, the trust would have controlled 60 percent of Rath stock. Unfortunately, considerable difficulty was encountered in designing a one person/one vote trust acceptable from

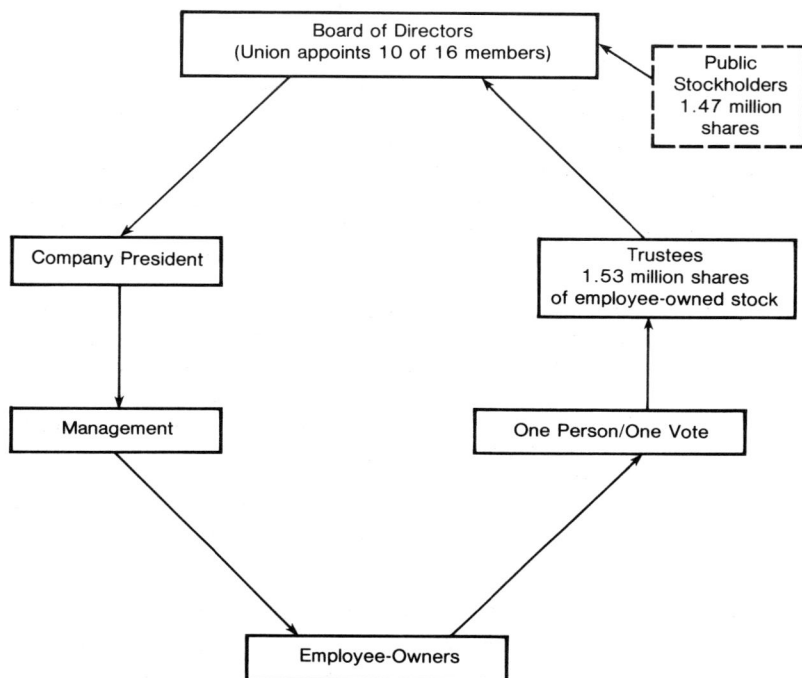

Figure 11.1 Structure of Employee Ownership

the standpoint of U.S. Department of Labor pension regulations, and Local P-46 and Rath expended over a year of effort and worked through several trusts and three different lawyers before arriving at the final plan. Because of the time it took to develop the ESOP, the plan was not finalized until many months after the stock purchase itself had been in process. Thus some employees took their shares early and sold them before the plan was formally instituted.

Nonetheless, a sufficient number of shares were committed to the Rath trust to provide for a 51 percent majority. Thus the trust was given the power to determine the composition of the board and yearly mandates for the board of directors, including the selection of company officials, establishment of corporate policies, and so on. The board, in turn, was to assume responsibility for directing Rath management in terms of goals, objectives and methods of operating of all spheres of activity. Figure 11.1 portrays the structure of this new and unique system intended to ensure both worker ownership and control.

COMMUNITY GAINS

The overall conversion to employee ownership provided important benefits to all parties concerned. The company received a substantial infusion of new capital. This has meant an annual contribution of approximately $5.7 million per year. The stock purchase itself generated an additional $3.6 million. The low-interest loan from EDA and the HUD monies received in the fall of 1980 generated an additional $7.5 million.

The city's major gain was the saving of 2,000 jobs with accompanying tax revenues and the retention of a substantial flow of funds to local businesses through Rath's payroll and company purchases within the area. Furthermore, the $3 million originally loaned to Rath for working capital by EDA created a locally controlled revolving loan fund. This fund has assisted numerous local businesses with low-interest loans.

For the union, the conversion to employee ownership provided greater equity in the making of sacrifices than would have been the case if they had conceded to the initial demands for concessions. Employees obtained a majority of the firm's stock and thereby gained a means for influencing the use of funds created as a result of their sacrifices. Through their control of the board of directors they were able to see to it that over $4 million of new capital was invested in badly needed plant modernization and improvement. Furthermore, the employees succeeded in not only salvaging the 800 jobs that existed when the idea of employee ownership was first conceived, but the company eventually increased employment to over 2,000 employees.

DEVELOPMENT OF WORKER PARTICIPATION
AND UNION-MANAGEMENT COOPERATION

Part of the significance of Rath as a case in employee ownership lies in its redefinition of the ESOP legal structure and its relationship to the board of directors, which was intentionally structured as a representative system capable of creating and maintaining a significant level of worker control. However, formal control through legal rights to vote in annual meetings and structure corporate policy were felt to be insufficient. When we joined the project as technical assistance resource persons in the fall of 1979, our interest was in facilitating the development of broader forms of union-management cooperation that would reduce the frustrated expectations of the new worker-owners through widespread worker participation—thereby avoiding the "time bomb" phenomenon discussed previously. We also believed that a more participative organizational structure would increase overall organizational effectiveness. Our thesis was that employee ownership would have to be accompanied by the introduction of a new pattern of industrial relations that would stimulate both productivity improvements and an enhanced quality of working life.

In February 1980, several months before the employee stock purchase actually started, a new system of union-management cooperation was initiated. Five top company executives and the bargaining committee of Local 46 began their efforts with the joint development of a charter statement formalizing the launching of this new parallel organizational structure at Rath, centering on a union-management steering committee. The charter document outlined both parties' commitment to collaborative problem-solving and clarified a set of values centering on employee motivation and satisfaction as well as company financial and productivity objectives. The charter also established some basic ground rules regarding the purview of the effort, and it created a steering committee to ensure both adherence to these ground rules and the achievements of goals and objectives. The charter strongly emphasized co-equal responsibilities in operating the steering committee and all other groups initiated by it for problem-solving purposes. (The charter and the committee were created distinct from the collective bargaining process, which continued to be the primary method of conflict resolution regarding grievances and the negotiation of new contracts.)

THREE-TIER STRUCTURE

During the first year of union-management cooperation, a three-tier structure emerged. First the steering committee itself began by evaluating problems and progress at the large Waterloo facility. Then the committee started to establish shop-floor Action Research Teams (ARTs). These groups, typically consisting of first-line supervisors, plant superintendents, division and departmental union stewards, and rank-and-file employees, met on a *nonpaid* volunteer basis to examine issues in their respective departments and divisions. Others were formed as ad hoc task forces to study and solve particular problems or design the layout and methods of operation for departments being modernized.

The steering committee assumed the role of launching new ARTs, monitoring their work, providing assistance where necessary, and, in general, functioned as a coordinating and sanctioning body for all cooperative activities.

A basic premise for forming the ARTs is that Rath employee-owners are skilled and responsible individuals committed to producing high-quality products in as efficient a manner as possible. Several hundred workers have contributed thousands of hours to the ARTs. The paragraphs below highlight the work of just a few of these teams.

Fresh packaged pork project.   Initiated by the steering committee to study the feasibility of beginning a new system of cutting and packaging fresh pork and to design the implementation plan, this team of thirteen managers and workers researched market possibilities, investigated alternative equipment and supplies, developed a productivity plan, and designed a self-managing work group structure to produce the new product line. This was the first of several projects initiated during the steering committee's first year.

Kill-and-cut departments. Night-shift ARTs were established in these two areas within the overall slaughter or abattoir division. The teams addressed safety, work flow, yields, and human resource issues—procedures, communication, and quality of work life. Over 150 different ideas for productivity improvement were generated, and the majority of these have been implemented.

Absenteeism team. An ART was formed to investigate the causes of high rates of absenteeism in the Waterloo facility. As a result of this group's efforts, a Peer Review Round Table was established for the plant. Composed of hourly employees and members of the management staff, this committee became the key body for administering corrective discipline and counseling to employees with severe absenteeism problems.

Energy conservation. A program to cut energy usage by 15 percent at Rath's Waterloo Plant was established in 1982.

Between the spring of 1980 and the fall of 1983, approximately twenty ART groups were formed, as well as a self-managing team spawned by the first effort. Teams were established in such areas as the maintenance department, bacon department, transportation, the office staff, the electronic data processing department, and so on. Within these larger groups, a number of mini-teams, sometimes referred to as line teams (referring to individual production lines within a department), were also established to carry out specific assignments and maintain ongoing communication with all departmental members.

It is important to note that the powers of ARTs increased considerably between 1980 and 1983. Initially, ARTs were, at best, accepted as

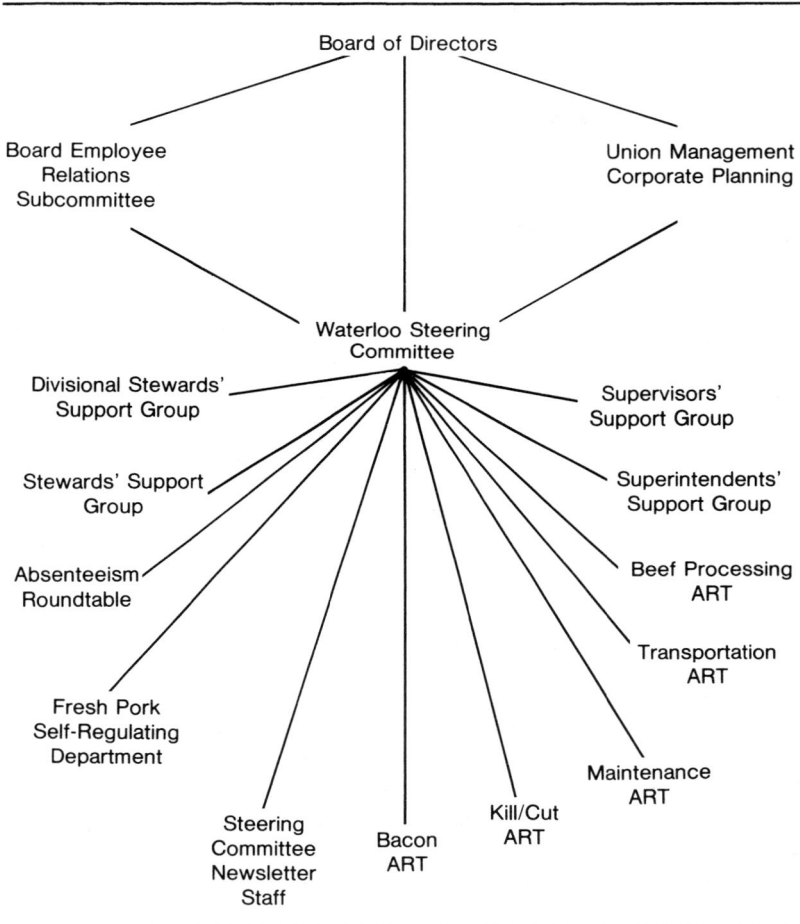

Figure 11.2 System of Union-Management Cooperation

advisory groups, and given little decision-making power. By winter of 1982, ARTs were delegated the authority to issue maintenance orders and make capital expenditures of up to $2,500.

A third level of participation in the ongoing affairs of Rath was also established—the steering committee. Two groups were formed with top management, top union, and board representation. One was the board Employee Relations Subcommittee, which was formed to listen to rank-and-file concerns and make management more responsive to these issues. The other group was the Union-Management Corporate Planning Group, which began meeting in 1980, as well as periodically for intensive two-to-three-day future planning-sessions. This group was formed to explore macro issues concerning new markets, critical financial problems, and possible scenarios two to five years into the future.

Figure 11.2 highlights the new structures of cooperation at Rath, the organizational mechanisms that parallel the line-staff and union-management roles of the past.

Gradually, a new vision developed as people experienced involvement in joint problem-solving. Evidence of this change can be seen in the results of an employee and management attitude survey that was administered to all employees at the fall 1982 "two-way communication" meetings. As can be seen from the data in Table 11.1, many employees at the Waterloo plant did in fact feel at that time that things changed positively with the advent of employee ownership.

In addition to the survey data, other benefits can also be cited as resulting from employee involvement. For example, the Absenteeism Round Table structure established in the kill-and-cut department and then adopted plantwide reduced unscheduled absenteeism from more than 8 percent to an average of 4 percent. The fresh packaged pork project became fully self-managing without the overhead expenses of a supervisor. The group members were trained to do various jobs and rotate as they desired in order to increase efficiency, reduce boredom, and so on. Thus the team took complete responsibility for such duties as ordering supplies, regulating production to meet sales requirements, setting up and cleaning the line, and even doing minor repairs.

Some employees also formed voluntary groups to perform functions not originally envisioned by the steering committee. One group, for example, devised the idea of a steering committee newsletter, entitled *The Input*, for which its members write, solicit contributions, edit, print, and distribute to all employees.

Employees also began coming in on their own time before work to discuss problems and seek information. Even laid-off workers would continue to attend and participate actively in ART strategy meetings—a phenomenon virtually unheard of in other U.S. firms with either employee-ownership or quality-of-working-life programs.

Employees began using their input on equipment purchasing decisions to ensure that their company was investing in "good stuff, and not the junk of the past," as one worker put it. Many workers even voluntarily began taking the initiative to take care of minor equipment problems or work around them instead of stopping and waiting for the arrival of maintenance as was common in the past.

People became more conscientious about cleaning up their own work areas, so much so that the clean-up crew was cut in half. In the truck

TABLE 11.1
Rath Employee Attitudes Regarding Change Since
Conversion to Employee Ownership (in percentages)

Since you first became a Rath Stockholder, do you feel . . .

	Decreased	Not Changed	Increased
(1) Your overall satisfaction with working for the company has . . .			
ART Member (N = 44)	11.3	29.5	59.1
Nonmember[a] (N = 155)	3.6	23.2	71.7
No ART[b] (N = 471)	9.6	25.9	64.6
(2) The amount of productive work you do has . . .			
ART Member (N = 44)	4.6	22.7	72.7
Nonmember (N = 151)	0.7	19.2	80.1
No ART (N = 459)	2.6	27.7	69.7
(3) In general, workers' say concerning their jobs has . . .			
ART Member (N = 44)	15.9	29.5	54.5
Nonmember (N = 151)	7.3	36.4	56.3
No ART (N = 452)	4.7	48.9	46.4
(4) In general workers' say in decisions in their own department has . . .			
ART Member (N = 40)	7.5	40.0	52.5
Nonmember (N = 150)	8.6	36.7	54.6
No ART (N = 455)	6.6	46.4	47.1
(5) In general, workers' say in overall policies of the company has . . .			
ART Member (N = 42)	9.6	40.0	66.6
Nonmember (N = 147)	8.9	32.7	58.5
No ART (N = 440)	6.6	46.4	47.1
(6) Communication between management and workers has . . .			
ART Member (N = 42)	16.7	26.2	61.9
Nonmember (N = 159)	6.1	34.2	59.2
No ART (N = 446)	9.0	38.1	52.9
(7) The amount of cooperation between employees has . . .			
ART Member (N = 42)	7.2	19.0	73.8
Nonmember (N = 146)	5.5	26.7	67.8
No ART (N = 445)	4.9	37.1	58.0
(8) Your concern with the financial results of the company has . . .			
ART Member (N = 43)	9.3	9.3	81.4
Nonmember (N = 152)	2.7	15.8	81.6
No ART (N = 449)	3.5	16.5	80.4

(continued)

TABLE 11.1 Continued

Since you first became a Rath Stockholder, do you feel . . .

	Decreased	Not Changed	Increased
(9) The confidence with which you can take your work problems to your supervisor has . . .			
ART Member (N = 41)	9.3	9.3	81.4
Nonmember (N = 150)	9.9	32.0	58.6
No ART (N = 452)	8.5	39.0	52.2
(10) The amount of information you receive on what needs to be done at Rath has . . .			
ART Member (N = 42)	7.1	19.0	81.0
Nonmember (N = 153)	3.9	23.5	72.6
No ART (N = 457)	4.3	24.1	71.5

a. Employees in departments with an ART who are not members.
b. Employees in departments without an ART.

repair shop, jobs were completed much faster and with better service than in the past—with fewer mechanics.

Pressure also developed from below for improved quality. The philosophy of the past, on occasion, was to "ship it no matter what." A new philosophy became to block bad material before it could leave the plant. When workers in one area saw their supervisor letting everything go out because he wanted to augment the department's yield figures, they went to the sales department to question this practice. The sales department, knowing the customer, backed the workers, and the supervisor gave in.

These bits and pieces of evidence are reflective of a synergy that emerged at Rath between the twin structures of ownership and shop-floor democracy.

ECONOMIC PERFORMANCE

Under employee ownership at Rath, many major economic improvements were made, especially through the sacrifices and cooperative problem-solving efforts of employee owners. The cost reductions alone, made as a result of wage and benefit deferrals, amounted to a savings of $26.8 million by the close of 1983; when the new equity capital gained through employee stock purchases is added to this figure, the total amounts to $30.4 million in cost savings over the four-year period.

In addition to the cost savings that accrued to the firm as a result of employee sacrifice in the form of wage and benefit deferrals, substantial savings were also achieved as a result of increased labor efficiency. Increased productivity resulted from the combination of greater motivation resulting from ownership, and the removal of impediments to efficiency through cooperative problem solving and accelerated capital investment. Heightened concern with the issue of productivity was stimulated by participation in both action research team activity and general employee meetings such as those that took place during the fall of 1981. At this time, general meetings were held with all Rath employees at Waterloo's downtown civic center. At these sessions, executives and union officials jointly described a plan for expanding into new markets and raising total sales in Rath's more profitable "processed meat" lines. More than 3,000 people attended these meetings, including both employees and members of their families. Follow-up meetings were then held in the Union Hall with each of Rath's many production departments that focused on productivity improvement. July 1981 was used as a base month and specific goals were established on a department-by-department basis. The short-term results of this program are illustrated in Table 11.2.

An examination of the overall picture also shows considerable gains in productivity between 1980 and 1983. Using 1978 as a base year, we find the average output per employee hour dropped by 11 percent in 1979, which was the difficult and anxious year preceding the buyout. In 1980, productivity improved 20 percent over 1978, and during 1981 it maintained at about the same level. The average productivity improvement over the base year was about 23 percent in 1982. Figures were not available for analysis of 1983.

In addition to the gains made in the area of productivity improvement, significant cost-savings also resulted from the energy-saving program and the introduction of new technology and improved plant layout design. The energy-saving effort, for example, succeeded in reaching its goals of reducing energy expenditures by 15 percent. In the area of plant improvements and new equipment, Rath's employee-owners invested $13.1 million between 1980 and 1983—a sum of $1 million in excess of what had been invested in the company in total during the fourteen years prior to the buyout. Another way to examine this issue is to compare the average ratio of depreciation with additions to plant and equipment before and after the buyout. Between 1966 and 1979, depreciation averaged 221 percent of additions to plant and equipment. After the buyout, through 1983, depreciation averaged 79 percent of additions and was as low as 34 percent in 1981, when Rath's

TABLE 11.2
Rath Production Comparison (average-$\overline{X}$-lbs. per hour): July 1981[a]
Versus Average for November 1981 Through January 1982

Department	July 1981	November 1981	December 1981	January 1982	Production Goal— Percentage Increase	Average ($\overline{X}$) Actual Percentage
Loin boning	97	81	97	129	16.3	5.6
Trim	263	313	274	299	0.0	12.3
Pix bone	262	328	334	342	4.7	27.7
Ham bone	162	149	172	171	13.0	1.2
Pork cure	617	955	972	898	15.6	52.6
Bacon	352	475	414	444	0.0	26.2
Dry salt	362	365	323	378	0.0	(2.0)
Beef	205	252	221	244	7.2	16.6
Smoking	292	347	375	333	16.6	20.4
Bacon slicing	276	196	205	188	10.0	11.6
Smoked meat	391	390	495	453	10.0	14.1
Roll sausage	569	737	673	718	29.0	24.7
Fresh ham	279	229	204	214	2.0	(33.7)
Dry sausage stuffing	143	153	158	147	34.0	6.8
Dry sausage packing	260	253	275	287	16.0	33.1
Canning	135	153	162	153	16.5	15.6
Loading	2194	2384	2380	2294	5.0	7.2

NOTE: $\overline{X}$ = average overall department productivity increase = 14.1 percent.
a. During the fall of 1981 "two-way communication" meetings at Rath, the July productivity figures were used as the base standard for establishing new productivity goals.

new employee-owners invested $5.5 million in new equipment and plant improvements. The lowest ratio during the fourteen years prior to the buyout was 89 percent in 1975.

In spite of many improvements made and cost-savings implemented after employee ownership, Rath's performance in terms of profitability was dismal after the first year. In 1980, everyone at Rath was overjoyed with a $3.3 million net profit. However, spirits crashed in 1981 with a net loss of $9.6 million on $465.4 million in sales. 1982 also yielded a loss; this time the figure was $6.5 million on $435 million in sales. This loss was later adjusted when the pension benefit guarantee corporation agreed to a termination plan for Rath's highly underfunded hourly and salaried pension plans. This was a difficult decision for Rath employees, but one that resulted in an adjusted net profit of $21.7 million for 1982 after the extraordinary credit created by the pension termination. At this point, the firm actually had a positive net worth, but this lasted for only a short time. By the close of 1983 Rath had again racked up another loss, in spite of the fact that all Rath employees took an additional $2.50 wage cut in February, and this time the loss amounted to nearly $13 million.

In light of the substantial cost-savings that resulted from the shift to employee ownership, it is difficult to understand why, after an initial profit, the firm maintained such consistent losses. One major factor in this situation was an increase after 1980 in the price of hogs. During 1981 and 1982 the price of hogs increased significantly to over one-third the price for which they sold in 1980 or, e.g., by $15 per hundredweight. This is quite significant considering the fact that raw materials are the most significant of all production costs in the pork-processing industry.

A second key factor involved with Rath's profitability problems was the emergence of many nonunion low-wage rate packers at approximately the same point at which Rath became employee owned. These firms continued to grow in size and number after the buyout, and frequently paid wages that ranged from the minimum standard to $5.00 per hour with few if any benefits. This situation made it very difficult for Rath to compete in the fresh meat/nonbrand-name product market. However, Rath could very likely have been quite competitive in the high-margin brand-name product market, and did indeed significantly increase its share of this market segment between 1980 and 1983 as a percentage of its total sales.

Unfortunately, a preliminary analysis at this point suggests that although Rath salespeople were increasing the firm's sales volume in the branded products, they may have also been simultaneously reducing the margins they made from these sales. This, therefore, points to a third and possibly major factor in Rath's demise, an unnecessary and fatal passing on of cost savings to the customer in the form of lower prices. For example, between 1980 and 1983 Rath's annual reports show that the firm gradually decreased the margin it charged over the reported cost of product sold. Specifically, the margins were 12.1, 9.3, 8.5, and 7.8 percent in 1980, 1981, 1982, and 1983, respectively. In contrast, Hormel, a somewhat (although not fully) comparable competitor, sold its products at margins of 18.7, 17.1, 16.8, and 19.5 percent during the same sequence of years while sustaining considerably higher labor costs.

Finally, a fourth key factor in Rath's demise was its continuing problem with lenders. The shift to employee ownership did little to improve the firm's credit standing. Thus, in order to sustain a sufficient cash flow to operate, Rath found it necessary to borrow funds at several points above prime with the loans secured against the company's accounts receivable. In fact, interest expense as a percent of net sales climbed continuously after the buyout. In 1980, interest amounted to only .2 percent of net sales. In 1981 it increased to .6 percent, in 1982 to .7, and by the close of 1983 interest expense was 1.1 percent of net sales. Even more debilitating were the credit policies followed by its principal lender. Substantial credit-line increases, for example, were offered only during low points in the business cycle or approved so late into peak periods that they were useless. Furthermore, when hog prices finally began to decline in 1983, providing a chance for profits, the lender insisted that the firm's president sign an agreement that set in motion a monthly reduction of the company's credit line from $20 million to as low as $10 million (given the lender's restrictive formulas for measuring receivables) over a six- to seven-month period. This policy was pushed even despite the protests of Price-Waterhouse, Rath's auditors, and by November 1, 1983, Rath management was forced to seek protection under Chapter 11 of the Federal Bankruptcy Code.

CONCLUSION

At the present time the fate of Rath Packing Company is unclear, and it is very likely that the company will face forced liquidation through Chapter 7 bankruptcy proceedings in the near future. The period since the declaration of Chapter 11 has been an extremely difficult one for all involved, and the strain of trying to maintain the firm with an ever-decreasing credit line has worn thin the cooperative labor-management relationship previously developed.

In spite of this disappointing ending, much can be learned from the pioneering experience at Rath. And whether or not Rath Packing survives as an employee-owned firm, it is clear that the pathbreaking work achieved there helped to spawn the emergence of other democratically owned and operated ESOP companies such as Atlas Chain in New Jersey, the O. and O. Superfresh Stores in Philadelphia, and the newly formed Bridgeport Brass Co. in Bridgeport, Connecticut.

Similar to the concern with protecting worker rights that motivated Local P-46 leaders to seek to develop a democratic ESOP and a system of worker representation that spanned from the boardroom to the shop floor, the democratic ESOP firms that have followed have sought to structure themselves along similar lines. By closely following both Rath's positive and negative experiences, they should be able to learn from the strengths as well as the weaknesses that emerged in the system. One thing that should be obviously clear is that the average worker, when provided with both a "piece of the action" and some measure of influence over life in the workplace, will respond with great dedication and sacrifice. From the considerable economic sacrifices made by Rath workers to the thousands of hours of volunteer time that they dedicated to shop-floor problem solving, it is apparent that the Rath effort at employee ownership tapped a reserve of hidden talent and motivation that simply never appears in the traditional enterprise in either good times or periods of economic crisis.

It is also clear from the Rath story that hard work and creativity on the part of employees cannot alone make a firm successful. Industry over capacity and turbulent market conditions can overshadow such achievements and are by no means susceptible to local influence. Similarly, blue-collar entrepreneurs, such as the people who stepped forward at Rath to make their contribution, need to be accepted more wholeheartedly by the banking community. The Waterloo group never succeeded in receiving the same support from lenders that would have been provided to the typical management takeover group. Finally, and in many ways most important, the Rath experience demonstrates that employees and union officials must learn to take on a stronger oversight role over management domains such as sales, marketing, and finance, which are unfamiliar but vital to a company's success. At Rath, employees would not have had to become managers themselves, but it would have helped if they had developed the ability to effectively supervise performance and policies in these areas. Unlike some popular management thought, which holds to the idea that workplace democracy can only result in difficulty because of too much worker intrusion into management, it appears from the Rath experience that a successful buyout must, over the long haul, involve comprehensive worker oversight in all areas of management. As former chief steward at Local P-46 Chuck Mueller said,

> Our problem at Rath was that we never really took control over the company. We let management do their thing unsupervised in sales and they drove us into the ground. If workers are going to really take over businesses, like we did at Rath, and make them run right, then they're going to have to jump in and take control over the whole company and not leave anything to blind faith in the ability of management.

NOTE

1. Mueller was the chief steward at Local P-46. Throughout the effort to develop an employee-ownership scheme, he and Taylor worked together as a team.

REFERENCES

BLASI, J. R. (1982) "The politics of ownership in the United States," in F. Heller et al. (eds.) The International Yearbook of Organizational Democracy. Chichester: John Wiley.

ROSS, W. (1980) "What happens when employees buy the company?" Fortune (June 2): 108-111.

U.S. Congress, House Small Business Committee, Subcommittee on Equity Capital (1979) Small Business Employee Ownership Act (H.R. 3056, May 8). Washington, DC: Government Printing Office.

U.S. General Accounting Office (1980) Employee Stock Ownership Plans: Who Benefits Most in Closely Held Companies? Washington, DC: Author.

WHYTE, W. F. (1978) "In support of voluntary employee ownership." Society 15, 6.

12

SAVING JOBS THROUGH
WORKER BUYOUTS

Warner Woodworth

In the fall of 1981, members of Local 736, United Auto Workers, gained legal control of a $100 million-a-year General Motors bearing plant in Clark, New Jersey. The firm immediately ranked as one of the largest worker-owned firms in the country, from then on to be known as Hyatt Clark Industries, Inc. (HCI).

Established as Hyatt Roller Bearing Company in the late nineteenth century, the plant was bought up by General Motors (GM) in 1916. An early executive at Hyatt, Alfred P. Slogan, went on to become president and chairman of the board at GM, and a major force in its gargantuan growth as a modern, bureaucratic corporation.

A key to the logic of GM's financial success lay in a marketing tactic that was founded upon the idea of planned obsolescence, that is, that the automobile would be built to last five years, after which it would be traded in on a new, larger, and more expensive vehicle. The technique epitomized GM success for half a century. However, this practice of trading for a new car carried GM straight into the wall of the 1973 oil embargo and OPEC with a resounding crash. The outcome was the death of the large, gas-guzzling monoliths Detroit had been manufacturing and the birth of a new consumer requirement that autos be smaller and fuel efficient. Japan and Europe became strong contenders for the American monopoly on auto production as GM's sales declined and the firm was forced by public demand to downsize its product.

New car design included, among other things, a switch from rear wheel construction to front-wheel drive. This meant the demise of the tapered roller bearings produced at Hyatt. GM decided the new spindle-bearing required for front-wheel technology would be manufactured at another bearing operation in Ohio, making the Clark facility obsolete. Over time, beginning in the late 1970s, Clark's products increasingly became less relevant to GM's scheme of things.

It was determined that retooling the plant would not be economically viable. Besides, there had been a long history of local union conflict with the parent corporation. GM perceived Local 736 leaders as militant, a fact attested to by a strike as recently as 1977. Wage rates and health and safety benefits were more expensive to the company at Clark than to sister facilities nearby. And costly grievances were numbered in the thousands. Gradually the noose that would ultimately cut off Hyatt from the GM system began to tighten. Beginning in late 1979, the number of layoffs grew incrementally—until by the summer of 1980 there were only 1200 workers left of 2400 the previous year.

THE BUYOUT STRATEGY

While the union leaders took a strongly declared stance to save jobs at Clark, GM was preparing to close the operation. The union studied alternative products that might be produced in the one-million square foot facility, sought new ideas from the rank and file, and offered more flexible work rules in the shop, all to no avail. In August 1980 GM announced the plant would close unless a suitable buyer could be found. The rationale for a closure was predictable: The Clark plant was unproductive, labor costs were high, the equipment and buildings were aging, and the new technology of front-wheel drive demanded different products. GM also initiated a shift to outside suppliers for parts rather than continue the monopoly from within. Known as outsourcing, this approach would theoretically allow the corporation to buy bearings or other parts from the cheapest high-quality supplier, thereby becoming more competitive in the marketplace.

In the ensuing months, as GM appeared adamant in its willingness to continue bearing operations and prospects for an outside buyer dimmed, the leaders of Local 736 began to seriously debate the potential of a worker buyout. They talked with local management, attorneys, consultants, and government officials about the options and steps necessary to take toward worker ownership. They reexamined rigidities in the labor contract and began to consider the possibility of changes that might contribute toward a more viable organization.

The local leaders sought rank-and-file approval to raise union dues to finance a feasibility study regarding worker ownership. The membership's reaction varied from solid support to doubt that worker ownership would succeed, disbelief that GM would really close, desire that benefits not be lost, as well as political opposition to the leadership in power. In a referendum in late 1980, the funding of a feasibility study was defeated by 16 votes out of over 1500 cast.

The next step came serendipitously from the opposition—Hyatt executives. Local managers began to fear for their jobs and sensed they might not be transferred elsewhere in the GM empire. So they resurrected the idea of worker ownership and passed out leaflets at the plant gates inviting workers to reconsider ownership as the sole means of surviving. A group of management and union leaders coalesced into the Hyatt Clark Job Preservation Committee to study and develop a method for gaining local control over their economic future.

The committee hired an attorney with extensive experience in worker ownership, tested out the idea of ownership with GM officials in Detroit, and moved toward financing the feasibility study. Instead of a vote, the committee sought $100 voluntary payroll deductions from anyone interested in a job in a new worker-owned firm in Clark. The result was impressive—over 1100 managers and workers at Hyatt contributed $125,000 and the study was completed by a major consulting firm, Arthur D. Little, Inc.

According to the study, the notion of a buyout looked promising if certain constraints were to prevail: Wages and benefits would become more competitive, the size of the work force would be diminished, GM would have to commit to purchase $100 million annually for the first three years of the new firm's existence, and management would have to achieve more effective results. Given this go-ahead, the Job Preservation Committee began to tackle the multitude of complex tasks it faced in bringing ownership to fruition.

Negotiations spun off in various directions. One thrust centered on discussions between local Hyatt people and GM headquarters, and issues ranged from payment of retirement pension benefits for GM employees (managers and workers) terminated by the closure to a sales contract for the next three years. There were internal struggles between the union and management leaders over a salary structure for supervisors and white-collar employees and a collective bargaining contract with Local 736. Still another task of the committee and its lawyer focused on the financial negotiations—a selling price from GM and loans from banks and other institutions.

During the summer of 1981, while these various challenges were all being confronted simultaneously, there was an aura of excitement, tension, fea: and conflict in the Clark plant. Meanwhile, all parties had to continue a fundamental chore: producing millions of bearings. There were fights over self-interests. Some managers ceased to be effective and looked for jobs elsewhere. Worker morale on the shop floor sagged as rumors spread regarding cuts in precious, hard-fought benefits. Time pressures built as General Motors' absolute deadline for closure approached.

On October 31, 1981 the purchase agreement was signed between General Motors Corporation and Hyatt Clark Industries, Inc., a 100 percent worker-owned enterprise. It consisted of 800 hourly workers and 150 salaried personnel.

LEGAL AND FINANCIAL STRUCTURE

The conversion to worker ownership at Hyatt Clark Industries, Inc., is based on a concept known as an Employee Stock Ownership Plan (ESOP), a highly innovative mechanism to finance businesses and obtain appealing tax incentives at the same time. The ESOP is a trust through which workers own the common stock of HCI. The ESOP bought a million shares of stock for $15 million that it borrowed from the Chemical Bank on a short-term basis. The plan is that this loan will soon be paid off by securing long-term, low-interest loans from a federal government Urban Development Action Grant, as well as lesser amounts from the Economic Development Administrations of Washington and New Jersey. Every year, as the interest and principal payments become due, the company will contribute sufficient dollars to the ESOP trust, as well as needed money to cover cash payments to retiring workers. Participants in the ESOP are credited with shares of the firm's stock to their individual accounts. Every time the ESOP makes a payment on the $15 million loan, the stock is released to the workers' accounts.

As these loans are paid off, the workers will become sole owners of HCI common stock. Both rank-and-file UAW members and HCI salaried/management personnel participate in the plan. After two months' service, all new employees also join the mandatory plan—thus all workers are also owners. Upon retiring, the worker will receive stock or a cash benefit reflecting his share of the equity that has accumulated over the years.

An important feature of the HCI legal structure is that the shares of stock are distributed according to seniority. This is not the case in most other U.S. ESOPs, in which management tends to have disproportionately more shares of stock, giving them ultimate control. Arriving at a better concept for HCI was not easy. Lawyers, consultants, and management agreed that shares of stock should be distributed on the basis of salary. In effect, this would mean management would obtain more shares. The union countered by demanding the stock be issued according to seniority. HCI is an old plant with a mature work force, so the result of this proposal is that the large numbers of workers with 25-30 years' seniority would receive the most stock. Ultimately, the union's logic ensued in which the concepts of worker justice and equality became relevant. After a bitter battle was fought just one day before the plant was to close, this issue was resolved in favor of seniority distribution. A more equitably based allocation of power and wealth had become a fundamental principle to the union, and in the end resulted in victory. The new president of the corporation now has the same equity as the new floor sweeper.

The trust has no assets other than the company's stock. At the outset, the trustee is the Midatlantic National Bank, which is subject to the direction of a three-member investment committee consisting of members of HCI's board of directors. In order to secure bank financing of the ESOP, UAW Local 736 agreed to defer voting rights of the workers' shares of stock for ten years. It is anticipated that by that date the ESOP will be fully owned by HCI workers and not be dependent on outside financing.

The total purchase price of the plant, equipment, and inventory was approximately $53 million. The lawyer representing the Job Preservation Committee successfully put together a package of financial resources that allows the workers to become owners without any direct, out-of-pocket costs as individuals. This was accomplished by obtaining $15 million from the Prudential Insurance Company to be paid back at the rate of $1 million a year for five years, and $2 million annually for the next five years. A second source was the Fidelity Commercial Bank, which agreed to supply up to $10 million for a revolving line of credit as working capital in the start-up of HCI. This amount can be renewed beyond the first year, if necessary. GM itself provided the remainder of the needed capital to launch HCI, through shares of HCI nonvoting preferred stock to GM—in effect deferring part of the purchase price. The projections are that GM assistance, as well as other lenders, will be fully paid off by 1991.

A NEW INDUSTRIAL RELATIONS

An important hurdle in the formation of the new company was the negotiating of a new collective bargaining agreement. Support for the lending institutions was contingent upon a new contract that promised a different future than the old union-GM relationship. Critical issues were fought through among a labor law firm brought in by the Job Preservation Committee, the union leaders and their own attorney, and the soon-to-be HCI labor relations management.

An implicit understanding had already been reached that concessions would have to be made in order to make HCI a more viable enterprise in the bearing market. In 1980-1981, unions all over America were negotiating labor contracts that cut wages, froze increases, and loosened up certain rigidities. The larger economic crisis was having a heavy impact on firms and unions generally, and HCI, in some respects, simply reflected the broad picture. However, it was more severe, as a plant closing was imminent, and worker ownership made it unique. At one level of worker ownership, it can be argued that union members were suddenly in the position of not making concessions to their employer, but rather to themselves. In another sense, these were not concessions at all, but rather exchanges. Reductions in wages and the like were merely traded for ultimate ownership of the stock, plant, equipment, and products of HCI. These views tended to undergird the contractual process.

In contrast to most negotiating efforts in the United States, at HCI the company's books were opened so that both parties were bargaining from a common data base. The reality of limited resources became apparent to all. With the evidence in front of them, union leaders ultimately agreed to give up 25 percent of the wage structure, restrict life insurance benefits, give up all personal paid holidays, and reduce their sick leaves. They also accepted a three-year no-strike clause to launch this new organizational experiment with lender confidence.

What this meant in specific numbers was that wages would drop from $12.00 per hour to $9.00 per hour. Again, it must be emphasized that this was not atypical of many other union concessions around the country. To survive and reopen as a worker-owned firm, $9.00 appeared to be operational within the budget constraints of the company. And it was still significantly above the prevailing wage structure in New Jersey, so HCI would continue to offer competitive incomes to potential new workers.

The union also bargained for an incentive system which, if desired productivity levels were achieved, would result in additional income to HCI workers. After much deliberation, an incentive program was agreed upon that consists of two types: one based on operating productivity of the entire work force as measured in sales per hourly labor dollar and paid each month; the other a semiannual profit-sharing program calculated according to overall HCI profitability. The emphasis is not on individual effort but the collective performance of all worker-owners together. Under such management, as the potential for productivity improvements is reached, members of Local 736 will actually become better off financially than they were with higher wages as employees of GM. And they come to have ownership and workers' control as well.

There were a number of other important agreements made during the process of negotiation that have made HCI a model for industrial relations in the United States. These include union representation on the board of directors, a variety of participative structures enabling workers to have a voice in company affairs, access to company financial records, posting of shop-floor performances as a feedback process throughout the plant, and so on. These will be discussed subsequently.

However, one additional piece of the labor contract must be mentioned—that of negotiating job classifications. The feasibility study had suggested that part of the declining competitive position of the Clark plant under GM was its increasingly cumbersome compartmentalization of work. Although such a system originally offered certain benefits in terms of making work assignments more explicit and creating jobs, over time these restrictions had reached the point of diminishing returns. By 1981, work in the plant was divided into 22 skilled-trade job classifications and 91 unskilled categories. The organization suffered from fragmentation, high labor costs, inefficient use of workers' time, and a demoralized supervisory group that felt unable to raise their departmental performance.

Both management and union agreed something needed to be done to breathe new vitality into this costly and archaic system. Eventually, a more simple and flexible approach was agreed to in which machine operators in certain areas of the factory would operate two or even three machines if it could be done efficiently and safely. It was felt that a small group could, with a certain degree of freedom, achieve more than a large group could under tight lines of demarcation. This was important, because the union knew it would have to become more productive with fewer workers to significantly diminish expensive labor costs. They also knew that the Hyatt work force was old (25 percent were over age 55) and that many would elect to retire early rather than start careers over with the risky new venture of worker ownership.

The new agreement reduced unskilled job classifications from 91 to 10, while that of the skilled work force shrunk from 22 to 9. In sum, from a bulky bureaucracy of over 100 categories, work at HCI is now divided into 19 distinct and more functional job classifications.

The creation of a new bargaining contract culminated in the creation of a document that, while not ideal, was agreeable to the union bargaining committee, company management, and the lenders. Clearly, the knowledge that the lenders were going to closely scrutinize the contract before final approval would be made for the capital to be advanced to the new company affected labor's openness to a significant departure from the old agreement. In the end, the new contract was completed in time to meet the strict deadline set by GM, and the membership of Local 736 ratified it.

Beyond bargaining with the union, it should also be mentioned that there were considerable negotiations affecting the salaried work force (clerical and managerial) that were going to affect the soon-to-start-up HCI. One set of issues among salaried personnel had to do with terminating their relationship with GM. When the big automobile manufacturer closes a plant, it has to pay a mutually satisfactory retirement benefit to eligible employees. In the Hyatt case, GM did not want to pay the "mutuals" to either the union or management, because it was assisting in the transition to a worker-owned firm in which people would still retain their jobs. With a backing of the International United Auto Workers, Local 736 won a satisfactory compromise affecting its members in which GM would add a fixed dollar amount to the selling price of Hyatt, and which would be paid back to GM over time. All eligible hourly workers would be fully covered by the mutual benefits.

But management had no union, and struggled for months against corporate headquarters in Detroit. As the reality of worker ownership appeared certain, the Job Preservation Committee found it could not successfully recruit management personnel from within the plant to stay on. A major reason given for this was that GM reneged on its termination benefit. Both GM and the union could see that without the talents and experience of the existing management, HCI could not successfully be launched. Finally, the salaried staff persuaded Detroit to lower its mutuals demand and an agreement similar to that of the union was created.

With that obstacle overcome, management and office recruits were quickly added to the new HCI organizational design. Another crucial controversy soon emerged, however, and this had to do with the salaries of management. Top executives had formulated a plan with roles, reporting relationships, and compensation. Upon learning that a number of key management posts involved higher salaries than under GM, a conflict with Local 736 erupted. The labor leaders argued that with the rank and file taking 25 percent pay cuts, it appeared management greed was moving executives in the opposite direction. Management countered that under GM, they were less accountable, with many of the decisions coming from Detroit. Now they bore full responsibility for a $100-million-a-year corporation.

However, the union pressed its charge of a contradiction, and eventually most top management salaries were negotiated downward. This was the first breakdown between management and the union in their joint effort to salvage the old factory and their jobs. It became a continued point of reference for subsequent union accusations that management could not be trusted. Over time, this breach of faith would widen. For every step toward a new trust, there were slippages backward.

PARTICIPATION AND CODETERMINATION

A major criticism of many of the emerging forms of worker ownership in the United States is their lack of any fundamental shift in power and participation of the union (Woodworth, 1981a). In the once widely heralded ESOP experiments of South Bend Lathe and Okonite, management took great pride in the fact that, under employee ownership, nothing had changed organizationally, that tradition was preserved in which management does the managing and workers do their jobs. This euphoria with the status quo ended in 1980 when both firms suffered long, costly strikes over a variety of issues—a central demand of which was more input into decision making.

One of the key thrusts of the union in its push to take over the Clark plant was that the new organization must not continue to function according to the GM mode. The notion was that management and workers at the new HCI must invent alternative organizational processes that would fundamentally alter the distribution of power, democratize the nature of work, and give the union some form of control at every level up to the top of the organization. A new bargaining contract and/or ownership on paper would not be sufficient to salvage and revitalize the aging Hyatt factory.

Thus it was, that from the first day of HCI's operation, a new, more humane climate was created. On that day, top management and union leaders met jointly in public meetings with groups of 120-150 workers and supervisors to discuss a philosophy of operations, goals, and expectations. Rotating teams went through the groups presenting overviews of expected sales, potential new products, manufacturing practices, and worker benefits. The results were electrifying, as workers of long seniority under GM began to catch a vision of creating their own business destiny. Long-held resentment and resistance to the distant Detroit bureaucracy subsided and a new esprit de corps was born. In the days that followed, evidence of these fresh winds of change was everywhere: Workers cleaned up their own areas rather than waiting for the custodial service; some took initiative and painted equipment that had been dirty and rusty for years; when the invitation went out for workers to propose the design of an HCI logo to symbolize the new firm, over 150 drawings were submitted, one of which became the official company logo.

A month later, as production seemed to be going well and the problems of starting up were eliminated, top management and union leaders spent several days studying alternative models of workplace democracy—Yugoslavia's self-management, the Israeli kibbutz, the Mondragón co-ops of Spain, Japanese quality circles, and German codetermination. They collectively reviewed cases of industrial democracy and/or worker ownership in the United States—Rath Packing Company, the Jamestown Labor-Management Committee, various quality of working life projects. Out of these days of study and deliberation came not only democratic education but a two-page "New Directions" statement—a philosophical commitment to innovation, worker involvement, and participative management.

Later in the month during a shutdown preceding Christmas vacation, approximately one hundred first-line supervisors, union bargaining committee members, middle management, engineers, and others were taken through a similar training session of several days each. Practical, experience-based skill training in consensual decision making, shop-floor meeting processes, and other concepts were introduced. A month

United Auto Workers Local 736   Hyatt Clark Industries

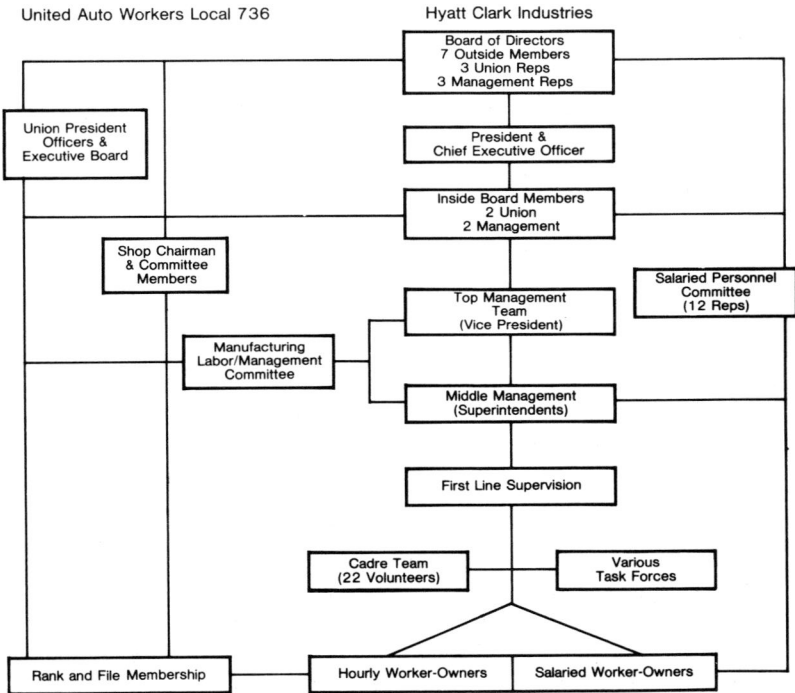

Figure 12.1 Worker Participation and Codetermination

later an HCI shop-floor foreman was coached in training so that he could begin taking small groups of office and manufacturing workers through a similar orientation. By now every worker and manager has had extensive exposure to alternative forms of worker ownership and industrial democracy.

Structurally, it was felt from the outset that philosophical under-standing and practical concepts were not sufficient to ensure a participative process at HCI. What was needed were mechanisms for operationalizing democracy in the organization. Figure 12.1 illustrates the tools created to institutionalize joint labor-management coopera-tion and problem solving. Participation essentially permeates the organizational climate of HCI, cutting across all departments, and flowing vertically from the shop floor to the board of directors. Some of the chief mechanisms that illustrate various kinds of labor/management codetermination are discussed in the paragraphs below.

BOARD OF DIRECTORS

In contrast to many U.S. ESOPs, the HCI case includes union representation at the top of the organization. During the many negotiations held in 1981 to design the form of worker ownership, leaders of Local 736 pressed for a union presence at the board level. After considerable debate, it was agreed that the union and management would each have equal board representation, with outside independent members in the majority.

The way it is currently structured, the directors include the company president, vice president of marketing, and vice president of operations. The union is represented by the president of Local 736, the chairperson of the bargaining committee, and a union-appointed professor who specializes in worker ownership. The attorney who was originally hired by the Job Preservation Committee to assist in designing the ESOP and arranging the financial package to buy the plant from GM is the board chairperson. He and six others, mostly prominent East Coast business executives, make up the seven independent board members outside the HCI plant itself.

The logic of having an impressive group of outsiders was that HCI would have more credibility with lenders, the state, suppliers, and customers. The idea of equal union and management representation implied a balancing of power from within the two constituent groups of the HCI operation. Over time, as the $53 million in financing is paid off, the workers will become de facto owners. After ten years they will be able to activate stockholder voting rights that have been deferred for now. Thus in 1991 the HCI work force will have pass-through voting rights allowing them to vote on any pertinent issues as stockholders, including the election of the board of directors. Full workers' control will at that point become a reality.

INSIDE DIRECTORS

A special office has been created at HCI to deal with major ongoing decisions facing the organization. Every Monday morning the president meets with the two union officials and two company vice-presidents who sit on the corporation's board of directors. These four officers study, evaluate, and make consensual decisions regarding various issues. Upon the meeting's conclusion, the president then communicates the inside directors' decisions, policies, and practices to his top management team later in the week. Thus the board has a direct, major impact on the company's ongoing work.

MANUFACTURING WORKS COUNCIL

To tackle the tough problems of revitalizing an aging manufacturing plant in such a way as to ensure improved productivity, quality, maintenance, and so on, a joint union-management works council has been formulated. It consists of union leaders, the vice-president of operations, and all plant superintendents. This group, trained in democratic methods and cooperative problem solving, is beginning to become an important institutional power resource, and plays a significant role in creating operations improvements.

SALARIED PERSONNEL COMMITTEE

Early into the process of worker-ownership, lower-level management, clerical, and technical people at HCI began to feel a growing sense of powerlessness. They complained of being neglected and left out of decisions affecting their jobs. Policies were being established for salaried personnel without office-worker representation. Thus in an attempt to broaden the scope of participation, a salaried council was established, consisting of twelve elected representatives ranging from engineers and supervisors to secretaries and computer experts. This committee meets monthly to air grievances, suggest solutions to problems, design and test out recommendations for top management, and so on.

SHOP-FLOOR TEAMS

At the production-worker level, groups of hourly workers, union representatives, and first-line supervisors meet weekly to evaluate their performance. As per the newly negotiated labor contract, there are bulletin boards with departmental productivity data posted throughout the factory. With the feedback from charts and graphs in hand, hourly work teams can now examine the causes of difficulties and generate concrete action plans for solving problems. The supervisors are better trained in meeting techniques, group skills, and listening and can coach the work team effectively. Workers have improved abilities to communicate as well as a more legitimate voice as owners of the operation. Hundreds of ideas have surfaced and been implemented through this important bottom-up approach to organizational effectiveness.

CADRE

A group of 22 workers were selected from nearly 100 volunteers among the work force to form a cadre of resources within HCI—a group of change agents. Their backgrounds included male and female, black and white, some with 25 years experience, some with only two years experience. Members include workers from assembly, inspection, maintenance, and the union shop committee.

This group received two weeks of intensive training in change theory, helping behaviors, cause-effect problem analysis, and other tools for providing technical assistance. Since then they have fanned out through the plant to provide additional kinds of help beyond their regular production or clerical jobs. A sampling of their activities:

- facilitate shop-floor team meetings in their own departments;
- sit in and observe the group process of another team when requested by a supervisor;
- initiate a study of quality circle techniques for possible implementation;
- begin working with techniques in developing ideas for new product development;
- establish a worker welfare committee to provide counseling;
- design an orientation/education process for newly hired workers as they come into HCI;
- launch an HCI newsletter, "Hyatt Highlights," to keep worker owners informed on business issues as well as to build worker solidarity.

Beyond the specific mechanisms highlighted above, there are a number of other examples of differing approaches and degrees of worker/management coparticipation at HCI. Several task forces have been organized to combat specific inefficiencies and problems of the past: One has attacked poor quality, another lack of maintenance, another emphasizes the high costs of energy and promotes conservation, another has made numerous improvements in HCI's shipping department—all the result of collaboration between managers and workers from different areas.

The company president has embarked on a serious program to achieve two-way communication with HCI's worker-owners. For instance, he holds a daily lunch meeting with a different group of workers each time to listen and respond to their concerns, test out new ideas that may be implemented, and generally "clear the air." The way it is conceived, this will enable each worker to achieve some face-to-face interaction with the chief executive officer of the company each year. In addition, at six-month intervals top management and union officials will hold mass meetings for the rank-and-file to inform them of overall progress, upcoming forecasts, and hear concerns from the people.

There have been structural changes designed to enhance communication and participation at HCI as well. For example, an extra level of supervision has been eliminated, the role of the assistant superintendents, so that foremen now report directly to three superintendents, who in turn report to the vice-president of operations. Flattening the pyramidal towers of GM management is a difficult task to achieve after forty years of being steeped in bureaucracy. However, the union is optimistic that more flexible institutional arrangements can be accomplished, ones that will loosen the flow of information, reduce expensive overhead costs for a top-heavy managerial elite, and make the firm more organizationally responsive.

LESSONS FROM THE HYATT MODEL

As HCI approaches its first year of operation as a worker-owned enterprise, several implications are beginning to emerge, and a few questions as well. The paragraphs that follow highlight major themes.

PRODUCTIVITY AND PROFITS

In spite of doubts and predictions to the contrary, HCI's performance under conditions of worker ownership are impressive. Within the first few weeks after it broke off from GM, the scrap rate had dropped from 11 percent to 7 percent, a significant savings in today's high-cost steel market. The amount of bearings shipped that reached quality standard increased from 98.5 percent to 99.2 percent. Over time, quality has been up and down, but with new technical training classes conducted by a master craftsman in the plant and gradual improvement of equipment maintenance, things look promising.

With respect to productivity, a clear pattern is beginning to develop. When calculating amount of sales shipped for the first half of the 1982 fiscal year, HCI has achieved $73,000 worth of sales per worker. This compares with $55,000 in 1981 and $50,000 in 1980. By controlling for inflation, in real dollars for 1982, thus far, output is up by 25 percent over 1981 and 1980. These are significant increases that portend the viability of HCI as the processes of worker-ownership and participation become further institutionalized.

CHANGING THE ORGANIZATIONAL CULTURE

After decades of existence as part of a GM division, a key challenge for HCI has been to develop its own organization and culture. Under GM, the plant operated as an extension of corporate headquarters in Detroit. A number of functions were handled by the bearing division offices in Ohio. Now worker-owned, HCI has had to build up its own engineering and quality control operations, and start from scratch with entirely new departments of sales and marketing, and accounting. A new vice president of finance has been hired and the structures are beginning to be established.

Local 736 leaders have been concerned from the outset that their company not be a duplicate of the GM model of the past. Some of the factors that were characteristics of the GM stereotype are clearly at odds with the desired values being implemented within HCI. The list below contrasts this shift in company culture.

Conditions Under GM	Potential of HCI
large	small
rigid	flexible
authoritarian	participative
elitist	egalitarian
work alienation	work identification
top-down control	shared control
organizationally fat	lean operation

To achieve this shift to some extent, the union has pushed to break the GM mold of having top-heavy, expensive management and salaried overhead costs. Today at HCI the salaried work force is small; supervisors are asked to not wear white shirts and ties, but rather to roll up their sleeves and work alongside factory operators. The elitist, private dining room of the GM top executive at Hyatt has been liberated and converted into a training room for workers as they become better educated about worker ownership and quality of work life efforts around the world. The parking lot, which used to have a designated area closest to the plant for management personnel, has now been opened to all workers equally—on the basis of first come, first served.

The old practices and attitudes are hard to abolish. For executives it means dropping the obsession with management prerogatives and beginning to view all HCI people as equals, as responsible co-owners in a radical new venture, rather than perceiving shop-floor workers as another breed. The politics of worker ownership requires a more progressive attitude. For the rank and file, changing the culture revolves around initiating constructive change, working for the collective good rather than individual job seniority rights, doing one's best in terms of quality and productivity, rather than just doing enough to not get in trouble with one's boss. Labor cannot merely react to management initiatives, but must become more responsible as comanagers. The union must engage in anticipatory problem solving to ensure company survival and the retention of jobs.

INTERNAL UNION POLITICS

The road to combating plant closures through conversion to worker ownership is not all sweetness and light for trade union membership. From the outset in HCI, the very notion was suspect. It is a different idea, implying new challenges and a redefinition of the union's role. Members of Local 736 were at first unwilling to increase membership dues to finance a feasibility study. A large number doubted GM would really close down, and therefore felt concessions would only play into Detroit's hand.

In May 1981 there was a bitterly contested election for union leadership. Those advocating worker ownership won in the end, but not without a lot of dissent and accusations. Until the day ownership was transferred from GM to HCI, opposing forces attempted to kill the effort. Union meetings were disrupted, equipment in the plant was vandalized, physical fights occurred among union members, appeals were made to the National Labor Relations Board, and the local labor leaders were taken to court under the pressure of an injunction leveled by dissidents to block the sale.

Local 736 leaders declared themselves to be good trade unionists, concerned about jobs and the welfare of their members. The counter charge was that they were pawns in the GM game, and had sold out to corporate America. In retrospect, after months of successful bearing manufacturing and the salvaging of 800 jobs at first, and then over 1,000 jobs, the union leaders are beginning to receive some hard-earned credit for the years of struggle they put into worker ownership. The militant activism of the past is still strong when it comes to fighting certain issues. But parallel operating tactics are also used now in contrast to the union's traditional militancy with GM. Now, there is considerable time and energy spent creating and shaping HCI's future, being proactive, rather than simply counterblasting GM decisions in the old days.

Local 736 leaders did much to win over the rank and file in the long run by an ongoing process of information sharing. Frequent membership meetings where rumors were dispelled and tough issues confronted head-on were held in the months preceding ownership. Almost weekly, leaflets were printed and distributed throughout the plant to inform members of lawyer recommendations, financial roadblocks, and the pluses and minuses of worker ownership. This up-front orientation activity did much to dispel myths, minimize speculation, and create an informed and confident rank and file response to the ownership idea.

But clearly, the concept of owning one's job is not about to be embraced by U.S. workers with open arms. It requires considerable debate, testing of assumptions and grass-roots struggle.

THE MOTIVES OF GENERAL MOTORS

An intriguing question in all this is why GM would agree to sell its bearing plant to an ESOP, and not only be willing to offer an acceptable purchase price, but go beyond—keeping the plant open until workers could prepare an ESOP, provide financing, legal help, accounting and other organizational assistance, and political clout in Washington. The most oft repeated explanation is superficial at best: GM simply wanted to unload a costly and inefficient plant by dumping it on one of the most militant trade union locals in the country.

With its considerable investment in the success of HCI and a three-year sales agreement, GM appears to have other interests that must be taken into account. One is the testing of the new strategy of out-sourcing, which in the end may make GM itself a smaller corporation, mostly engaged in automobile assembly rather than having such a large, monolithic structure of suppliers, vendors, and assembly all under the same corporate umbrella. Another motive appears to have been to make HCI a public illustration of what workers can do for themselves to become more effective, save jobs, and preserve the industrial sector of the economy. A third motive includes a willingness to accept concessions by Local 736, which GM hoped might loosen up the International UAW for upcoming bargaining talks on an industrywide basis.

Finally, GM has been concerned about its public relations of recent years. Other plant closings by Ford and Chrysler were the subject of much citizen criticism across the country—from government officials, the media, and the public. Severe economic conditions already in existence in New Jersey made the Clark plant closing extremely vulnerable to public outcry. When U.S. Steel Corporation refused to even consider selling its Youngstown, Ohio, mill to its steelworkers and instead brought in the bulldozers, the sentiment across America was outrage. GM wanted not only to offer its workers a deal, but to actively support this experiment in grass-roots capitalism.

Down the road, the HCI effort may become a prototype for other GM disinvestments, providing a solid option to lost jobs and economic dislocation.

THE ROLE OF THE INTERNATIONAL UNION

There have always been conflicts between local unions attempting a worker buyout and the international union officers. In the South Bend Lathe case, regional and international officers of the United Steelworkers actively attempted to block the ESOP effort. In the case of Jeanette Sheet Glass in Pennsylvania, the local glass and ceramic workers union decided to withdraw from the international because top union officials did not assist in the shift to a worker-owned firm (Woodworth, 1981b).

So it was in the case of HCI that there was considerable division within the United Auto Workers. UAW headquarters had a long history of dissatisfaction with Local 736. Like GM, it viewed the union as militant, perhaps even leftist beyond the political stance of the international. In a sense, Local 736 had been on the bad side of the international for years as a kind of troublesome, maverick local.

The worker-buyout idea seemed preposterous. Union officers in Detroit were not prepared to cope with the concept, nor had they any in-house expertise to evaluate what was occurring in New Jersey. In addition, they were fearful that concessions at HCI would set a precedent for upcoming industrywide contract talks. Thus the international officially took a position that Local 736 efforts were simply a regional experiment about which Detroit had no opinion.

Union leaders in Clark interpreted this as a rejection of the idea, and in subsequent convention discussions clearly felt criticized for attempting the buyout. However, their response tended to be one of asking what else should be done. If ownership was not the best answer, what was?

To date, relations between the two UAW levels seem to be improving. On the one hand, the international has completed current bargaining talks with GM and achieved some assurances regarding lost jobs and plant closings. On the other, HCI is proving to be a viable option for job retention of UAW members. Currently, the international has opened its doors to Local 736 leaders and begun to seek their advice on a UAW official position regarding plant closings and worker ownership. The idea is now being discussed of having an international union task force on feasibility studies, ESOP designs, and technical assistance to local UAW groups facing a shutdown.

CONCLUSION

The struggle to successfully design and finance a worker-buyout in the United States is a tough, complex process. As evidenced in the HCI case, it requires dynamic labor leadership, a harnessing of highly skilled expertise that is congruent with working-class values, and the unifying of various and often conflicting interest groups in achieving an overarching goal—the salvaging of a closing business and the right to a job.

For ownership to be successful, the HCI experience also suggests the need for power parity. In other words, there is a need for competent managers and professional knowledge of finance, marketing, and so on, coupled with a strong union, capable of long-range vision and codetermination in economic decisions.

Whether or not the HCI effort will ultimately be financially viable, only the future will tell. Clearly, workers and management must push hard to accomplish two things: (1) They must build a solid performance system for assuring high-quality bearings, so that GM will extend its contract and other business with Ford and Chrysler will materialize in the coming years; and (2) they must become highly innovative in seeking and developing new products that will free HCI from its vulnerable dependency on the automobile industry. The extent to which this can be done within the constraints of an aging facility, outdated equipment in poor repair, with entrenched GM autocratic methods of managing remains to be seen.

In spite of the long-term uncertainty, the fact of the matter is that over a thousand jobs have been saved in New Jersey. A payroll that would have disappeared has gone into the work force of HCI—over $25 million in wages, benefits, and productivity incentive annually. Perhaps there is an alternative to a large-scale economic concentration in corporate America. So far, HCI shows promise of having the internal strength and high commitment necessary to effect a grass-roots political economy. Whether or not such a radical system can survive in the larger capitalistic context is still in question.

REFERENCES

ROSEN, C. (1981) Employee Ownership: Issues, Resources, and Legislation. Arlington, VA: National Center for Employee Ownership.

WOODWORTH, W. (1981a) "Forms of employee ownership and workers' control." Sociology of Work and Occupations 8, 2: 195-200.

———(1981b) "The emergence of economic democracy in the United States." Economic Analysis and Workers' Management 15, 2: 207-218.

PART IV

PROSPECTS FOR REVITALIZATION
Barriers and Resources

Recent problems of declining productivity, increased foreign competition, and record numbers of bankruptcies and industrial shutdowns have made local strategy development to cope with these dilemmas imperative. In the past, the response to such problems was typically to view them as inevitable, the "natural" consequences of market forces over which workers and communities had no control.

Since the late 1970s this fatalistic perspective has been changing, with area labor-management committees and worker ownership as two dominant forms of reaction. However, the process of achieving regional revitalization is a complex activity that requires not only grass-roots cooperation but the mobilization of various resources.

The first chapter in this section is a contribution by Joel Cutcher-Gershenfeld, who carefully describes some of the policy issues facing area labor-management cooperation at the federal, state, and local levels. He makes a case not merely for funding programs, but for other kinds of support as well, including training programs and technical assistance. An important development in the future will be the shift beyond area labor-management cooperative projects to a broader redefinition of the general field of industrial relations.

Next, in Chapter 14, Corey Rosen of the National Center for Employee Ownership discusses the financing of employee ownership. In it he debunks the common assumption that workers themselves must personally come up with the funds to purchase their ailing businesses. Instead, leveraged strategies are utilized to obtain needed finances, much like the financing methods of conventionally owned businesses. Rosen reviews a number of cases in which workers become owners and illustrates a variety of successful avenues to financing such enterprises. The cumulative evidence is impressive: The mix of lenders includes government, banks, unions, and individuals—and the list is growing.

Joseph Blasi of Harvard University next addresses some implications of worker participation and ownership in light of the National Labor

Relations Act. Although federal laws were created to protect both workers' and managers' rights, such legislation was based on assumptions of an adversarial relationship. Currently, with workers participating in management decisions and, in many cases, obtaining stock and board of directors membership, a number of questions begin to emerge. Blasi describes the constraints of labor policy on these institutional problems and suggests needed changes by government, management, and unions to create more successful economic organizations in the future.

The book concludes with a synthesizing chapter by Meek, Whyte, and Woodworth that attempts to delineate key theoretical and practical elements about community economic reindustrialization. They suggest that new social inventions and interventions are characteristic of successful attempts to transform regional economic problems into grassroots change. The implications of these new strategies for a more participative behavioral science and for future action research are described.

13

POLICY STRATEGIES FOR LABOR-MANAGEMENT COOPERATION

Joel Cutcher-Gershenfeld

Diversity and flexibility are concurrently key virtues of area labor-management committees (ALMCs) and major liabilities. Because ALMCs have been established in a variety of communities, their ability to adapt to different circumstances has been essential to their operation. But this same adaptability has complicated and sometimes hamstrung the development of public policy—particularly federal policy—in support of joint community efforts.

Among the over fifty area labor-management committees that have been established in recent decades, the only clear common denominators are the equal involvement of top local union and employer leaders, as well as a shared commitment to improving the local labor-management climate. Beyond that, ALMCs vary considerably in their focus, structure, funding, and impact. Moreover, most individual ALMCs have evolved substantially since their inception. Although this diversity across ALMCs and flexibility within individual ALMCs bespeaks a certain responsiveness to constituent needs, it confuses efforts to specify the policy significance of ALMCs. Specifically, it is hard to pin down ALMCs as serving a particular purpose because their significance lies in being vehicles for addressing a wide range of policies.

The development of local, state, and federal public policies toward such committees is further complicated by the context within which they operate. In one sense they are but one of a number of forms of labor-management cooperation. Although distinct, their fate is inextricably tied to public policy on other forms of joint activity. In another sense all of these cooperative activities are embedded in the larger context of overall U.S. public policy on labor-management relations. Before enumerating policy recommendations with respect to ALMCs, it is important to trace the evolution of policy to date (paying special attention to the implications of ALMCs' diversity and flexibility) and to position such history in the larger context of current U.S. labor-management relations.

At the outset, it should be noted that policy positions on community labor-management cooperation are often treated as synonymous with public funding for ALMCs. Although this may be the most important expression of policy support and an appropriate focus in tracing the history of policy on ALMCs, it is not the only policy option. As such, the analysis and policy recommendations in this chapter will address symbolic and programatic support, as well as financial support.[1] It should also be noted that the history of public policy on community-based joint committees has not proceeded in a linear, orderly fashion. Rather, it has occurred in fits and starts in response to a diversity of pressures and on the basis of a variety of conceptions of community-level cooperation. A central assumption of this chapter is that policy development need not continue in this way—that it is possible for policy to be driven by a coherent view of ALMCs, which is grounded in actual ALMC experience and embedded in a larger vision of the role of labor-management cooperation in the U.S. industrial relations system. This may be an overly optimistic assumption about the way government operates, but it is a foundation on which policy discussions should rest.

VEHICLES FOR CHANGE

ALMCs have played a critical role in (1) promoting local investment, (2) attracting new industry, (3) establishing joint work-site committees,

(4) fostering employee involvement, (5) reducing strikes, (6) training displaced workers, (7) upgrading the skills of employed individuals, (8) helping to resolve labor-management disputes, (9) sponsoring joint conferences and educational events, and (10) generally establishing a community-level network of communications among unions, companies, and other local institutions. These activities fall under a range of policy umbrellas, including: economic development, full employment, productivity, the quality of work life, training, community education, and industrial peace. From a policy perspective, ALMCs can neither be constrained as a tool for addressing just one of these issues, nor can they be touted as a panacea, able to address all of these issues in every community.

Unfortunately for ALMCs, public policy at federal, state, and local levels is typically oriented toward addressing specific issues. It is far more rare to see policy support for the establishment of new instrumentalities or vehicles for addressing a range of issues. Tracing the history of governmental support for ALMCs does, in fact, reveal this initial preference of an issue-oriented focus. It also reveals that, once a committee has demonstrated a distinctive competence that is not invasive to existing power structures, it is possible for a joint committee to be supported as an important end in itself.[2]

At the federal level, the issue-oriented focus has been a constant source of frustration to local labor, management, and community leaders which may be endemic to policy in other areas as well. In the early 1970s, when an increasing number of committees were established, it was common for them to contact U.S. representatives and senators only to be told, as one early traveler recalled, "You've got the greatest idea since sliced bread, but we don't have a program to fit" (Gershenfeld and Costanzo, 1982: 13). When funds were forthcoming it involved ALMCs positioning themelves in relation to single issues. John Popular (1985), executive director of the National Association of ALMCs, has characterized this focus as a macro perspective on a micro problem.

Two early sources of financial support for ALMCs—the Economic Development Administration (EDA) of the U.S. Department of Commerce and the Appalachian Regional Commission (ARC)—both stressed economic and community development. These two organizations provided support to a variety of ALMCs from 1973 to 1980 (EDA) and from 1976 to 1981 (ARC) for those of their activities that fell within these domains. Job training has been stressed in financial support from the Comprehensive Employment and Training Act (CETA), which supported such ALMC activities from 1976 to 1982. Its successor, the Job Training Partnership Act (which will be discussed in more detail below), added to job training a specific focus on economic dislocation. For a few years after its establishment in 1975, the National Center on Productivity and the Quality of Working Life did provide symbolic and limited financial support to ALMC efforts oriented toward employee involvement. Until 1976, the technical assistance arm of the Federal Mediation and Conciliation Service (FMCS) was also an aid to ALMCs, but primarily by virtue of ALMCs' capacity to contribute to the prevention of disputes. Federal mediators are still involved in many ALMCs, but not as part of any formal policy initiatives.

Although it is true that key individuals within the above mentioned administrative agencies have conceived of their support to ALMCs as reaching beyond a particular legislative mandate, the support still positioned ALMCs as but one of many vehicles to accomplish a specific legislative end. As those legislative priorities shifted, so did that particular stream of support.

The first major effort to provide federal support for ALMCs per se was initiated after a special election in 1976 in western New York that brought Jamestown Mayor Stanley Lundine to the U.S. House of Representatives. He came to Congress full of the experience of cofounding a highly successful ALMC. Senators Javits, Mathias, and many other colleagues in Congress joined in the effort to extend this experience. Even then, however, the support was positioned in the context of bills to promote "full employment and economic stabilization."

Both the Human Resources Development Act (HRDA) of 1976 (HR 14269 and S 3783) and the HRDA of 1977 (HR 2596 and S 533) included language on labor-management cooperation, but neither generated sufficient support to be reported out of committee. The two HRDAs suffered, in part, because of their broad focus. The elements of the bills that addressed employment issues were perceived as a threat to the then pending Humphrey-Hawkins legislation[3] and the provisions on unemployment compensation and capital formation raised a complex set of questions in testimony (Leone et al., 1982: 221-233).

It was, however, these two bills that provided a foundation for the National Labor-Management Cooperation Act (LMCA), which became law in 1978. The LMCA, although not limited to ALMCs, still represented the first federal legislation to center on labor-management cooperation rather than on any specific outcome. At the time, the adaptability of the ALMC model was supported by Arthur F. Burns, former chairman of the Federal Reserve Bank, who stated in a letter to Representative Lundine:

The significant merit of your Jamestown program, I believe, is its flexibility to adapt to community efforts in solving particular regional problems [cited in Leone et al., 1982: 223].

Although the Act allows for higher levels of funding, it has only been supported at annual levels of approximately $1 to $2 million since its first allocation in 1980. About 20 percent or less of that money has been distributed among ALMCs, with the rest going to public and private work-site committees, state-level or regional efforts, and one national initiative. In its administration the Act has been constrained to providing short-term seed support for innovative efforts. The effect of this policy has, on the one hand, ensured the distribution of limited funds over a variety of initiatives. It has, on the other hand, forced established ALMCs into the familiar position of trying to fit into a specific policy (innovative programs) rather than being supported in their generic role as multipurpose vehicles.

In addition to the Labor-Management Cooperation Act, which is administered by FMCS, the only other current source of federal financial support for ALMCs has been generated under the Jobs Training Partnership Act. Part of this support has come through an administrative ruling allowing local Private Industry Councils (PICs) to fund ALMCs, which had been constrained under initial JTPA rules and regulations. The actual provision of support is dependent on the priorities and, frankly, politics at the local PIC-level. Also, under the JTPA the U.S. Department of Labor made a set of grants of $25,000 each to nine ALMCs for training-related programs. Although critical to the effectiveness of many ALMCs, both sources of JTPA support are issue-focused on training and economic dislocation. Moreover, these allocations are viewed by some JTPA officials as diverting resources outside of their central mandate. This source is not likely to be stable in the future, dependent as it is on being a priority for the Secretary of Labor or other high officials.

Beyond sources of financial support, both the U.S. Department of Labor (DOL) and the Federal Mediation and Conciliation Service (FMCS) have recently lent symbolic and programmatic support to community-level cooperation. The U.S. DOL has funded the most complete study to date on the operation of ALMCs (Leone et al., 1982) and it has provided information on ALMCs as part of its Bureau of Labor-Management Relations and Cooperative Programs. Both the DOL and FMCS have worked with the National Association of ALMCs in two national conferences on labor-management cooperation (focusing on many levels, including that of the community) and the

DOL has worked with individual ALMCs in regional conferences. Finally, as noted earlier, a number of federal mediators continue to play active roles in various ALMCs. Unlike the pattern in federal funding, these efforts have treated ALMCs as multipurpose organizations.

Also, by contrast with the pattern of federal funding, state-level policy—where it exists—has emphasized general support for ALMCs. This has been true in Kentucky since 1979, where a statewide joint council was established partly to provide technical assistance in the establishment of new ALMCs. This advisory council has also overseen the provision of some direct financial support for ALMCs from state government. The notion of state-level support was also embodied in a Michigan Quality of Work Life Council initiative (funded via an FMCS demonstration grant) to help establish new ALMCs. This effort is now part of the goals of a statewide coalition of Michigan ALMCs and has received modest support from state coffers. The most far reaching state effort is currently taking place in Pennsylvania. Recent legislation has provided close to $.5 million for general operational support for the state's nine ALMCs, administered by the tripartite MILRITE Committee (Making Labor and Industry Right in Today's Economy). Similar state-level efforts are under consideration in Wisconsin, New York, Ohio, and elsewhere.[4]

These state efforts, far more than federal initiatives, reflect ALMC support that allows for individual diversity and flexibility. This is not to say that ALMC advocates have not been pressured to position themselves in relation to specific issues; they have—especially in the area of job training. Rather, it indicates that there are a handful of states in which there are a sufficient number of ALMCs to argue that, in and of themselves they represent a network deserving of support. This is only possible, of course, after the ALMCs become well-established at the local level.

In-kind and direct financial support from city and county government is at the heart of the operation of many ALMCs. Although this form of support is most likely to conceive of ALMCs as multipurpose institutions, it is often contingent on ALMCs first establishing themselves with separate funding. Unlike the situation at the state and at the federal level, the diversity across ALMCs is not an issue at the local level as there is only one ALMC under consideration. On the other hand, the evolving focus of an ALMC may be an issue with respect to local policy. Thus ALMCs receiving local support must attend to shifting local priorities and, sometimes, are subject to local pressure to continue with whatever focus was the basis for initial public support.

A few ALMCs exist as quasi-governmental agencies. This is the case in Toledo and Louisville, where the cooperative efforts are line items in city budgets and the boards include equal numbers of labor, management, and community members. Both of these ALMCs were funded by local government in the late 1940s in support of their efforts to minimize contentious labor-management relations. Although these two organizations have also made contributions in the areas of economic development and employee involvement, their focus remains on grievance and contract disputes. At this point, the mission and the mechanisms supporting these two ALMCs are fairly well-institutionalized.[5]

For most U.S. ALMCs industrial retention and expansion have been the common formative issues. They have emerged from a diagnosis that an adversarial labor-management climate is sapping local economic vitality. Interestingly, however, local policy-makers have not necessarily made direct or in-kind support contingent on specific results in addressing these issues. More typically, in the cases in which a local public official has played a key role in the formation of the committee, there have been various in-kind forms of support made available at the outset. Regardless of the role of local public officials, which is a policy decision itself, the provision of financial support has generally been forthcoming after the committee has established itself in the community. This choice was forced upon many committees in the late 1970s as various sources of issue-specific federal support dried up. In some cases, as in Lock Haven, Pennsylvania, local government did not fill the gap. For other ALMCs, such as those in Buffalo, Cumberland, and Jamestown the city and county funds were forthcoming. For the many ALMCs formed in recent years, often with federal or state seed money, the issue is sure to arise again.

In reviewing the history of public policy on ALMCs it becomes clear that the policy has been driven by field innovation, rather than the reverse. Before there could be state or federal support (albeit seed support) specifically designated for ALMCs, a critical mass of committees were needed to demonstrate the viability of the concept. At the local level, ALMCs often have to demonstrate similar viability before city and county government will provide support. The process has been particularly complex because each ALMC reflects its community and its problems at a given time. This has made for confusion over just what are ALMCs. In retrospect this path may seem inevitable for a new social institution. For the labor and management leaders involved in ALMCs, however, it has been especially frustrating because they typically see themselves as solving problems rather than creating new institutions.

A LARGER CONTEXT

The preceding overview of public policy on community labor-management committees has stressed the evolution from issue-oriented support to institutional support. However, it is important to see this evolution in the context of increasing public attention to labor-management cooperation at the work-site, regional, industry, or national levels. It is especially at the work-site level that cooperation has received attention for improving productivity, quality, dignity, and participation.[6] Though this is again an issue-oriented focus, it has added a certain legitimacy to the general notion of cooperation.

With the increased public attention to labor-management cooperation has also come controversy. Work-site innovations such as participation teams, autonomous work-groups, and gain sharing have been introduced to unionized settings not only jointly, but often unilaterally by management. In these situations the effect, and sometimes the intent, has been to undercut unions. The debate within unions and across the labor movement over work-site cooperation has weakened policy efforts to support ALMCs. It is a debate that has been carried out in the strongest terms. Some have argued that participation is necessarily "phoney" and incompatible with trade-unionism, while others have argued that it represents a critical new component of union activity (U.S. Department of Labor, 1984: 54; Watts, 1982: 1).

In a general sense, the lack of a consensus on work-site cooperation has fostered the perception that community-based cooperation is also controversial. It was this sort of perception that led the Michigan state AFL-CIO to take a cautious rather than aggressive position in support of efforts to assemble a statewide program of support for ALMCs. Support for national legislation was marked by an ambivalence similar to that experienced at the state level in Michigan (Leone et al., 1982: 321). In order for legislation to be passed in support of ALMCs in Pennsylvania, the distinction between ALMCs and QWL had to be made especially clear. In truth, some ALMCs are actively involved in promoting work-site cooperation—where there is strong local support for that activity.[7] From a policy perspective, it is important to see ALMCs as having the potential to support work-site cooperation at the option of local participants, but not as necessarily impelled to pursue that program area.

There is an organic link that does exist between ALMCs and QWL, but it can only be seen by stepping out of the domain of labor-management cooperation and into the larger field of industrial relations. This is a field that demonstrated remarkable stability for over three decades following the Supreme Court's upholding of the National Labor Relations Act and its subsequent institutionalization under the National War Labor Board. At the heart of the system was what Dunlop (1958) characterized as a "web of rules" codified via collective bargaining. Lower-level shop-floor relations and upper-level policy and strategy decisions were all defined in relation to the middle level of collective bargaining.

In recent years we have seen important, often contradictory changes occurring at all three levels of the system. These include: aggressive union-avoidance strategies conducted by employers; increasing union access to information and decisions on corporate investment; a declining proportion of the work force represented by unions; a relative upsurge in employee and community ownership; the emergence of ALMCs; a rush of concession agreements with a legacy of shifts in the substance and process of bargaining; increasing world competitive pressures calling for more flexibility and higher quality; shifts toward more participative management styles; the wholesale elimination of entire tiers of management; increasing employee and union concern with job security, health and safety, employee involvement, gain sharing, and pension investment; and a blurring of the line between blue- and white-collar work. A number of scholars have argued that the net effect of the changes will be a transition in industrial relations comparable to the upsurge of public-sector unionism in the 1960s or the rise of industrial unionism in the 1930s or the dominance of business unionism that emerged in the late 1890s (see Kochan et al., 1984; Kochan and Piore, 1985).

Although it is too soon to tell what a new industrial relations system will look like, we can be fairly sure that any emerging system will only be stable to the extent that it features no major inconsistencies across all three levels of the system (Kochan and Piore, 1985). It is also clear that at least some of the recent changes in the traditional industrial relations system will endure.[8] From a theoretical and a practical point of view, the resolution of the transition is likely to require situating an evolving "web of rules" within the larger process of reconceiving a web of relationships.[9] During the current transition and, probably, wherever we end up, it is also likely that existing legislation on labor-management relations will be found wanting.

Distinctions between exempt and nonexempt employees do not reflect the current reality, in which blue-collar workers are increasingly involved in daily and sometimes long-term policy decisions while white-collar workers are feeling increasingly vulnerable. Distinctions among permissive, mandatory, and prohibited subjects of bargaining are inconsistent with union involvement in policy decisions and the larger community impacts of labor-management relations.[10] Similarly, the legislative intent to protect the right of employees to organize is violated in both the administration and the structure of U.S. labor laws. The duty of fair representation and a plethora of workplace legislation have vastly complicated grievance and arbitration procedures. Meanwhile, there is growing tension between the concept of exclusive representation and the structure of various participative initiatives (Schmidman and Keller, 1984).

The resolution of these and other policy contradictions—which can only follow the reemergence of a coherent industrial relations system—will have profound implications for the future of ALMCs. Indeed, to the extent that these community-based organizations emerge as enduring features of a new U.S. industrial relations system, it is likely that public policy will contemplate areawide joint discussions of corporate investment, a higher community profile at the bargaining table, and joint administration of local training and education programs.

POLICY STRATEGIES

Summing up the status of public and private cooperative efforts, U.S. Congressman Stanley Lundine recently observed,

> Faced with intense international competition, United States industry has taken steps to establish joint labor-management committees, improve product quality, and enhance the quality of work life. Some progress is being made in breaking down old adversarial relations. Employee ownership, especially using the tax benefits of ESOPs, is a more significant factor in economic revitalization. Nonetheless, government support of these efforts and their widespread use still lags behind what is needed and is an area where progress must continue [Lundine, 1985].

In the short term it is unlikely that dramatic shifts will occur in U.S. public policy regarding overall labor-management relations. In fact, if this is truly a period of transition, major policy initiatives should await the emergence of a stable system. In the meantime, there is an opportunity to choose among three transitional policy paths that would achieve the sort of progress that Lundine has called for.[11]

Along the first path, which most U.S. industrial relations legislation has followed, the policy is forged in a crisis, often involving the triumph of one set of advocates over another. The National Labor Relations Act, the Landrum Griffen Act, the Occupational Health and Safety Act, and many public sector laws proceeded, to some extent, along these lines—encountering inevitable resistance in their implementation. Under this plan the government shuns formal commitment to institution building although a crisis may force it into that role.

An alternative path was urged by John R. Commons via the Wisconsin Industrial Commission and followed, to an extent, in the passage of the Railway Labor Act and in the establishment of the Economic Alliance for Michigan. This path features policy development based on field research and experimentation, followed by a degree of consensus-building among practitioners—so as to inform and temper policy debates. In both of these options policy follows practice.

The third possibility involves the passage of legislation that is intended to introduce a social program or particular innovation. Though some poverty programs and other social programs are built on this model, there is little precedent for government-driven innovation in industrial relations.[12] Thus, if left to the first two paths, the choice is clear. Underlying the following set of recommendations is a conviction that the middle path—viewing this transition as a time for encouraging innovation—will ultimately produce the most effective and enduring policies.

Assuming, then, the middle path, the range of policy options at local, state, and federal government include: (1) observation and evaluation; (2) supporting pronouncements; (3) information diffusion (written and via public events); (4) application of cooperative principles for government employees; (5) technical assistance; (6) seed funding; (7) institutional support; and (8) alignment of overall legislative environment. Each will be considered below.

For policymakers at all levels, the implication of a commitment to encouraging social innovation is to learn more about current ALMC activities and to develop mechanisms to track emergent efforts. For many, discovering what is happening in their own back yard may be a pleasant surprise. Some movement has occurred in this direction, but it could be better coordinated and supplemented by a series of comprehensive evaluation studies.[13]

To the extent that the new information indicates that useful contributions are being made by joint efforts, policymakers at all three levels have the option to employ what Teddy Roosevelt called the "bully pulpit." Government has the capacity to legitimize the risks that many labor and management leaders take through public statements and other symbolic actions.[14] If, however, the goal becomes the fostering of broader experimentation with cooperative efforts at the community level, then a further-reaching agenda is required.

At the local level, the first option beyond public pronouncements lies in the involvement of labor and management in various issue-specific initiatives. For example, in Battle Creek, Michigan, there is no ALMC, but the board of a local economic development effort includes more than token labor representation. Labor's distinctive competence in getting out the vote on a critical development issue, its ability to ensure smooth construction on new industrial sites, and its daily contributions on various subcommittees has helped the local economy in ways that would not have been possible with the traditional preponderance of employers on such community efforts. Similar contributions are possible via labor and management roles on the boards of community colleges and in the administration of local training programs.

For local policymakers there is a related issue of government's role as an employer. Many of the principles of employee involvement that might be fostered in the private sector have also been used effectively in the public sector.[15] To the extent that local government urges labor-management cooperation, it also has the opportunity to practice what it preaches.

Finally, for local politicians and other community leaders there is the option of working with labor and management leaders in the direct establishment of an ALMC. There is precedence for such individuals initiating the first meetings of companies and unions to discuss the concept or for their joining existing explorations. The process can involve behind-the-scenes meetings and public forums (such as a recent conference on labor-management cooperation sponsored by the Greater Cleveland Roundtable, 1984). Creating such an organization allows the advantages of the issue-specific involvement noted earlier, ensures support for public-sector employee involvement efforts, and allows for a forum adaptable to future needs. This is, of course, an option that is constrained by the level of unionization in a community. It also raises a set of questions as to the appropriate level and forms of governmental support. In-kind support is common, though line-item budget support from cities and counties is emerging as perhaps the most significant institutional form of support for ALMCs.

States are just beginning to take up a mantle of responsibility for providing financial support to ALMC efforts. It is too soon to evaluate the effect of such programs, though it is not too soon to put in place thorough mechanisms for conducting such an evaluation. Also, the largest effort to date—Pennsylvania's legislation—is just focused on the provision of direct financial support. States should consider also the provision of technical assistance in organization of new ALMCs and the forging of formal networks among a state's ALMCs. Both would maximize local innovation.

Apart from making a commitment to local ALMC efforts, states can and do benefit from labor-management involvement in issue-specific and more generic state-level committees. The Maryland Productivity Center, the Michigan Quality of Work Life Council, the Economic Alliance for Michigan, the Northeast Labor-Management Center, and the Work in Northeast Ohio Council are a few examples of such entities. Like ALMCs, these organizations are highly dependent on institutional support and certain to evolve over time—possibly revealing that institutional adaptability may be an important feature of an emergent industrial relations system.

At the federal level the vast majority of top-level, joint efforts have been centered on specific issues. Assuming that these will continue to be established as appropriate (though the case has been made for a much higher labor profile in such forums), a broad array of policy options exist that will strengthen local labor-management cooperation. Current efforts by the U.S. Department of Labor to gather and disseminate information on joint initiatives should be maintained and possibly expanded into a decentralized computer information service. Similarly, regional and national conferences should be continued for their content and their ceremonial importance. However, the focus of these activities should move even further toward linking with state and community joint initiatives to maximize local innovation and contribute to the stability of local efforts.

Without changing the current legal structure, a few administrative changes could still have a significant impact. The FMCS could renew its commitment to technical assistance in the organization of new ALMCs— at least it could give symbolic and programmatic support to mediators filling this role. Also, the funding and administration of the National Labor-Management Cooperation Act has fallen short of its potential. The eighteen-month maximum time period for support may be appropriate for other types of joint initiatives, but it is insufficient as a start-up period for ALMCs (Leone et al., 1982: 243). The Act does not fully address the needs of established ALMCs and congressional appropriations have been inadequate. The Act is a good first step toward encouraging innovation in a time of transition, but a vigorous commitment to labor-management cooperation would recognize that joint efforts above the work-site level have medium-term as well as short-term needs for institutional support.

Ultimately, U.S. public policy in the coming years would best be driven by a shared vision that this is a time of change, accompanied by a willingness to tolerate variation. In this context ALMCs are but one of many new institutional arrangements that should be watched and encouraged. During this period the task is to seek a diversity of models and to ensure that these many efforts are given the institutional support that cannot be generated by labor and management when their own world is in transition.

In time, broad debates over federal policy will come to the fore. Even if they are contentious, they should at least be informed by a commitment that is possible only now—a commitment to social innovation. Only then will we be fairly well assured of institutionalizing a new, stable web of labor-management relationships.

NOTES

1. Also, although the analysis will touch on a variety of value-laden issues, it is only the final section of this chapter (dealing with policy recommendations) that is intended to be normative.

2. For a fuller development of this issue, see Trist (1978).

3. William Wimpisinger, president of the International Association of Machinists (IAM), commented in testimony before the Senate:

> Quite frankly there is a real fear among many people in the labor movement that some Congressmen or Senators may view the Human Resources and Development Act as a way of bypassing H.R. 50, the Humphrey-Hawkins Balanced Growth and Full Employment Bill. If these two bills were being considered on an either/or basis, we would much prefer the Humphrey-Hawkins solution [cited in Leone et al., 1982: 225].

4. Certain state-level and regional efforts should also be distinguished from these ALMC efforts, including the Northeast Labor-Management Center, the Maryland Productivity Center, and the Work in Northeast Ohio Council. These and similar efforts do involve joint labor-management committees, but their focus is not primarily at the community level. Their contributions are important and many of the policy recommendations noted in this chapter do complement these issue-specific, state-level efforts.

5. Indeed, the local legislation supporting the Toledo organization specifies that the ALMC shall prepare a budget in October and the city council shall appropriate the money, which has occurred consistently for nearly four decades.

6. This literature ranges from the many volumes and articles on Japanese management styles to over a dozen years of popular and academic publications on the quality of work life and employee involvement.

7. Although the Jamestown ALMC is, perhaps, best known for its work in this area, ALMCs in Lansing, Cumberland, Buffalo, Philadelphia, Kenosha, and elsewhere have all

made major commitments to such activity. On the other hand, ALMCs such as the one in Jackson, Michigan, have explicitly avoided this area and others have not given it high priority.

8. Current research efforts in the industrial relations section of MIT's Sloan School are centered on this very issue, with support from the U.S. Department of Labor.

9. This theme is developed more fully in Joel Cutcher-Gershenfeld (1985).

10. These issues are addressed in Kochan (1985).

11. The explicit conception of this as a time of experimentation is based on a conversation with Thomas A. Kochan (January 31, 1985).

12. Industrial relations legislation in other countries has followed this model. The recent French legislation mandating a quality-circle form of shop-floor participation might be one example of this.

13. The National Labor-Management Cooperation Act does feature an evaluation component that has recently shifted from self-evaluation to an independent model. It will be important to trace the results from this effort.

14. In discussing the legitimizing role of the federal government in the area of labor-management cooperation, Sidney Harmon, then president of Harmon Industries and undersecretary of commerce, observed:

> There is one federal government. When it sanctions, in whatever terms it chooses to do so, and dignifies this kind of process through some kind of legislation or other, in my judgment, it tends to coalesce the many interests that exist in a rather diffused way. And it does indeed do that very important thing: it legitimizes, it sanctifies the process. I don't know anybody that could do it as well [cited in Leone et al., 1982: 228].

15. At the federal level the U.S. Navy, the U.S. Department of Labor, and other agencies have made significant progress in this area. New York State has the largest state-level initiative (initially receiving over $1 million) and there are literally dozens of municipalities that have efforts under way.

REFERENCES

CUTCHER-GERSHENFELD, J. (1985) "Reconceiving the web of labor-management relations." Presented at the annual spring meetings of the Industrial Relations Research Association, Detroit, April 17-19.

DUNLOP, J. T. (1958) Industrial Relations System. New York: Holt, Rinehart & Winston.

GERSHENFELD, J. E. and C. G. COSTANZO (1982) A Decade of Change: The Ten Year Report of the Jamestown ALMC. Jamestown, NY: JALMC.

Greater Cleveland Roundtable (1984) Cleveland's Economic Survival: The Need for Labor and Management Cooperation. Cleveland: Author.

IAM (1982) "Quality of work life programs." IAM Research Report 10 (Winter/Spring).

KOCHAN, T. ed. (1985) Challenges and Choices Facing American Labor. Cambridge: MIT Press.

———R. B. McKERSIE, and H. KATZ (1984) "U.S. industrial relations in transition: a summary report." Presented at the annual meetings of the Industrial Relations Research Association, Dallas, December 28-30.

KOCHAN, T. and M. J. PIORE (1985) "U.S. industrial relations in transition," in T. Kochan (ed.) Challenges and Choices Facing American Labor. Cambridge: MIT Press.

LEONE, R. D., M. F. ELEEY, D. WATKINS, and J. GERSHENFELD (1982) The Operation of Area Labor-Management Committees. Washington, DC: U.S. Department of Labor.

LUNDINE, S. (1985) Personal communication (January,.

POPULAR , J. (1985) Personal communication (January).

SCHMIDMAN, J. and K. KELLER (1984) "Employee participation plans as section 8(9) (2) violations." Labor Law Journal (December).

TRIST, E. (1978) "Directions of hope: recent innovations interconnecting organizational, industrial, community, and personal development." John Madge Memorial Lecture, Glasgow, Scotland, November 3.

U.S. Department of Labor (1984) Labor-Management Cooperation: Perspectives from the Labor Movement. Washington, DC: Government Printing Office.

WATTS, G. E. (1982) "QWL and the union: an opportunity or a threat." Work Life Review 1 (November).

14

FINANCING EMPLOYEE OWNERSHIP

Corey Rosen

In the past decade, we estimate that 60-65 companies have been bought by their employees to avert a shutdown. This represents only 1 percent of all employee-ownership cases, but has generated more than its share of publicity and public interest. Of these 60-65 cases, 4 have closed and 2 are in Chapter 11. An estimated 50,000 jobs have been saved in the process, although roughly 3,000 jobs have been involved in the unsuccessful cases. Other parts of this book discuss how these buyouts were organized and are now working; this chapter will focus exclusively on how financing for employee buyouts is arranged.

For most people, the idea that a group of employees can arrange to raise a sum ranging from a few hundred thousand to several hundred million dollars to buy a troubled company seems, on the face of it, impossible. Visions are conjured of employees about to face unemployment raiding their savings, mortgaging their homes, cajoling their relatives, and otherwise scraping together whatever is needed to buy their company and finance its almost inevitably needed modernization.

At a cost per job ranging from several thousand to $100,000, raising these sums seems totally impractical, not to mention outrageously risky. In fact, a few early employee-buyouts were financed this way. In general, however, much more sophisticated and less demanding strategies are used, generally financing most of the buyout through "leveraging"— borrowing money against the assets of the firm to be purchased. The employees risk as little of their own capital as possible, relying on the income their new company will generate to repay the loan. The employees are no different in this model from any other investor. Investors rarely use current assets or other income-generating properties to buy companies. Instead, they borrow the money on the assumption that if the investment is economically sound, it will pay for itself. Banks, and other creditors, make their own judgment and decide whether the investment makes sense for them as well.

Of course, in a buyout situation, potential creditors are naturally going to be more reticent about loaning money. The business in question is usually in some trouble or the employees would not be trying to buy it. Moreover, the idea of employees owning, and sometimes controlling, a company is still foreign and thus suspicious to many lenders (although the problem has eased considerably in recent years).

Finally, before a buyout can even be considered, a feasibility study must be performed. These can cost from $10,000 to $100,000 or more. Workers faced with imminent unemployment are naturally reluctant to put up the $100 or more per worker that these studies often cost, especially because results are uncertain. For all these reasons, approximately one of three employee buyouts has received some form of government assistance, whether it was a direct loan, a loan guarantee, or a feasibility study-grant.

Within these very general boundaries, employee-buyout financial packages differ considerably from case to case. Several basic types have emerged, however, and the rest of this chapter describes these, illustrating each with some of the more prominent examples.

EARLY BUYOUTS:
WORKER/COMMUNITY DIRECT PURCHASES

In early to mid-1970s, several employee buyouts were accomplished primarily by employees or employees and community members digging into their pockets and buying their company. In 1971, for instance,

Northwest Industries decided that its Chicago and Northwestern Railroad did not fit into the company's future plans. The railroad had been losing money, and prospects for a turnaround seemed slim. No other company seemed very interested either, so C&NW's management made a proposal to buy their own firm. Northwest agreed to sell the company for $19 million, plus the assumption of the railroad's debts by the new owners. This represented a bargain to the new owners, and produced a tax loss that Northwest could use to offset its other profits. Some 300,000 shares of new stock were issued, and 1,000 employees bought 25 percent of them, raising $3.6 million in equity (a down payment used to obtain loans for the remainder, which was to be paid over twenty years). The remaining shares were available for later purchase. Approximately 12,000 employees did not buy shares on the first offering, and the majority of the 1,000 who did were management people.

In a short time, C&NW had turned around completely and more employees decided to buy stock, although no more than a minority of workers were ever actually owners. Those who were owners often became very wealthy—C&NW stock's value increased to seventy times its original value by 1980. Eventually, C&NW made its stock available to the public as a means of raising additional capital, and now only about 35 percent of the total is owned by employees (McClaughry, 1974).

The C&NW approach resulted in a highly skewed form of employee ownership, albeit one that was financially very successful for the company and the employees who did buy shares. A smaller company, the Library Bureau, used a similar approach, but was able to obtain much broader participation in ownership not only by employees but by members of the community as well. The Library Bureau was formed in 1876 in Herkimer, New York. A manufacturer of library furniture, the company had come to be owned by Sperry-Rand, which found its "low-tech" line of business fit poorly with its own plans. Although the Library Bureau was earning a modest return on investment, Sperry-Rand sought higher returns elsewhere. In 1976, it announced its plans to liquidate the plant, leaving 178 people out of work. Since unemployment in the area was already pushing 14 percent, community leaders and local management turned to employee ownership as a possible solution. After what were sometimes heated negotiations with Sperry-Rand (aided by some political clout exercised by former Senator Jacob Javits), it agreed to sell, but left the employees only a short time in which to raise the funds. Remarkably, within days, initial subscriptions for the sale of stock raised almost $2 million. Most employees and many community residents bought shares, often digging deep into personal savings or arranging bank loans using personal property as collateral. Private and federal loans totaling $4 million were also arranged, and employees were able to save their jobs.

Since the buyout, the company has had a somewhat checkered experience. Over the years, managers bought more shares and took clear controlling interest. An ESOP (more about these later) was established to spread ownership more broadly to employees, and it now owns just under a majority of the shares. Initially, the company had considerable financial success, but in the past three out of five years it had lost money. The direct ownership of shares by employees and community residents proved to be an unworkable way to maintain employee ownership because employees and residents had no reason to hold on to their stock indefinitely. Because they could sell it at financially opportune times, management was able to buy a greater share of the company. Still, the company now employs 250 people and believes that with the improvement in the economy its fortunes will rise (Hammer and Stern, 1980).

TRUST-BASED OWNERSHIP

Several problems emerged with these and other efforts by employees to buy stock directly. Often, management bought most of the stock. Employees could and did sell their stock at will. Perhaps most importantly, employees had to use savings or borrow money on funds that had already been taxed, which created a major financial burden and risk for them. Although a few companies were purchased this way, many more buyout efforts attempting this approach never got off the ground.

A more practical approach relied on using some form of a trust to accomplish the buyout. Here, the employees form a new company and have it establish a special trust to hold their ownership interest. That interest cannot be sold until the employee leaves the company, and distribution rules ensure that most or all employees will share ownership on an equitable basis. All of the buyouts discussed in the remainder of this chapter use this approach, relying either on an ESOP or worker cooperative approach.

The simpler of the two techniques is the worker cooperative. In a cooperative, only workers can be owners. Each owner owns one and only one share, and has one vote. Cooperatives can, and often do, however, hire nonowning workers, sometimes on a temporary, seasonal, or part-time basis. In some cooperatives, the share values reflect the actual worth of the company, and if the company is successful each worker's share can become very costly. This can make it difficult to recruit new worker-owners, because they may be reluctant to purchase a share from a departing owner in order to become a cooperative member. Often, cooperatives allow the new worker-owner to buy the share over time, relying on payroll deductions to circumvent this problem.

Another approach, based on the successful Mondragón cooperatives in Spain, keeps the ownership fee artificially low. In fact, the fee is not considered a form of ownership, but is rather a basis for membership in the cooperative organization. In this approach, no one is thought to "own" the organization individually. It is owned collectively by all the people who work there at a given time. Individual cooperative members have full and equal rights of control but no direct equity in the asset value of the company (Ellerman, 1980).

In either case, cooperatives are uniquely allowed to shelter part of their profits from income taxation. Any net earnings a cooperative makes can be allocated, in part or in their entirety, to members as a "patronage refund" (another term for profit sharing). Earnings returned to members are not taxable to the cooperative, even if up to 80 percent of these earnings are kept in an account for the worker-owner until he or she leaves the company. Most cooperatives use some form of this internal account system, drawing temporarily on the account funds for working capital (money for inventory and other short-term needs). Individual workers must pay tax on any patronage refunds (even if they do not actually receive more than 20 percent of them in cash). In the internal account systems, the amounts kept in the accounts are allowed to accumulate, with modest interest, until the worker leaves. If there are net losses in a year, however, these are subtracted from the worker's account.

Cooperatives are appealing to many worker-ownership advocates because their structure ensures democratic, one-person, one-vote control. In most cooperatives, the board of directors is composed of employees, and many decisions are made by consensus by various employee committees.

Cooperatives do not have some of the financing advantages of ESOPs, however, and the very requirements that ensure they are 100 percent worker owned and controlled make them difficult to use for some buyouts in which lenders or investors are hesitant about this degree of workplace democracy. In recent years, however, a few companies (not buyouts) have been able to combine the financing advantages of an ESOP with the governance structure of a cooperative. ESOPs are considerably more complicated and flexible. In an ESOP, a company establishes a special trust fund designed to invest primarily in the stock of the company. To enable the trust to make this investment, the company either contributes cash to buy its own stock, issues new or treasury stock (stock issued before but not sold to anyone yet) to the trust, or has the trust borrow money to buy stock. Regardless of how the trust acquires the stock, it must allocate the shares, with some exceptions, to all full-time employees with at least one year's service who are over the age of 21, or, occasionally, over age 25. What "allocation" means is that each participating employee has an account established in the trust in which that employee's shares are held (not physically, of course, but on the trust's books). The allocations can be made on the basis of relative compensation (if you make $40,000 and I make $20,000, you get twice as much) or some more equitable formula. Most companies use relative compensation as the basis, although it is often adjusted somewhat to produce a less skewed distribution. A few companies allocate shares equally.

Allocated shares are kept in the employee's account until he or she leaves the company or retires. The employee is "vested." Vesting is the process by which employees receive a gradually increasing right to whatever is in their account. Usually, an employee becomes 30 percent vested after three years, with the percentage gradually increasing to 100 percent after ten years. If an employee leaves after five years on this schedule, for instance, he or she would usually receive 50 percent of the stock in the account. If the employee has actually purchased any shares, however, as is the case in about 2 percent of all ESOPs, vesting is immediate for those shares. Unvested shares are returned to the trust and distributed to everyone else.

Once the employee has the stock, he or she can require the company to repurchase it at its fair market value, as determined by a stock exchange if the company is publicly traded or by an appraisal expert if it is not. Most employees do just that; if not, the company does have a right of first refusal to ensure that shares do not fall into unfriendly hands. In companies that are substantially employee-owned (80 percent or more), employees can be required to take the cash value of their stock.

Employees are not taxed on the stock while it is in the trust, and are taxed only minimally when they receive the stock. In publicly traded companies, employees must be able to vote their shares on all issues; in privately held companies (almost all buyouts), they are only required to be allowed to vote on a few select major issues. About 60 percent of all majority employee-owned companies, including buyouts, voluntarily pass through full voting rights. It is important to note that in most cases this is on a one-share, one-vote basis, not one-person, one-vote. A few ESOPs, such as Rath Packing and Weirton Steel, however, do use a one-person, one-vote system.

ESOPs have many tax advantages, which largely explains why 7,000 companies have installed them since the first ESOP laws were passed in 1974. At the time they set up their ESOPs 98 percent of the ESOP companies were profitable. One particular ESOP advantage, however, has a special relevance to buyouts—the ability of ESOPs to borrow money and have the company repay it in pretax (cheaper) dollars. And, equally important, with the passage of the 1984 Tax Act banks were allowed to exclude 50 percent of their income from loans to ESOPs.

Esops AND CO-OPS IN ACTION: DIFFERENT WAYS TO RAISE MONEY

DOING IT ALONE: LITTLE OR NO GOVERNMENT INVOLVEMENT

Most people are surprised to learn that the majority of buyouts have had little or no government assistance. We estimate, however, that this is true for about two-thirds of all buyouts. A major case of this approach, the Weirton Steel case, illustrates how this can happen.

Weirton Steel is certainly the most celebrated case of an employee buyout, and it was basically privately financed. On March 2, 1982, National Steel announced that it would no longer invest in its giant Weirton Steel division. Employees could either buy it, or National would turn Weirton into a small finishing plant. Considering that Weirton employed 7,000 people in a town of about 30,000, and that the town was located in a state with the highest unemployment rate in the country, the buyout seemed a necessary choice. Moreover, it seemed a practical choice. Weirton had made a very small return for National on its investment, just not as much as National could make elsewhere.

The employees and management formed a joint study committee and hired the most expensive, well-known firms they could find to conduct the feasibility study and negotiate the purchase price (on the theory that these firms would enhance the effort's credibility).

Part of the over $2 million in legal fees the buyout eventually cost was paid for by the state, but most was raised by the employees through their union, by local people (with everything from telethons to bake sales), and from other private sources. The feasibility study showed Weirton could make it, provided employees took a 32 percent cut in pay and benefits (later, after negotiations brought down the sale price, this figure was cut to about 18 percent). Extended negotiations with National began.

National wanted to sell the mill, not close it. The costs of closing the mill had been estimated at over $700 million, mostly for pension payouts and severance pay. If Weirton could stay open for five years—National's legal period of obligation to continue to pick up pension costs— National could save a fortune. Weirton's negotiators used that fact to secure a sale price for the company of $66 million, which was only 22 percent of the book value of the assets. That meant that Weirton had instant equity (the remaining 78 percent, which lenders could sell if Weirton defaulted). In addition, Weirton would pay $200 million much of it at long-term favorable rates of interest, for the current excess of National's assets over liabilities in Weirton (that is, the value of the supplies and products minus debts the company had). In addition, Weirton would have to raise about $700 million more over ten years to modernize.

All this money, with a few small exceptions, was secured from major banks and National itself. Weirton also set up an ESOP to finance the debt, and, like Hyatt, has a minority of employee representatives on its board. After five years, however, employees can begin to elect the remainder of the board and, in the meantime, can vote on any issue stockholders would normally vote on, mostly on a one-person, one-vote basis.[1]

After losing money in the two years between the 1982 announcement and the January 1984 completion of the buyout, Weirton reported a profit in the first quarter of 1984, and rehired 600 workers.

This pattern has been repeated in a number of smaller buyouts. In many, but not all cases, the former owner helps with the financing. Even if there is no direct involvement by the former owner, however, a cooperative attitude is essential. The former owner must supply financial information, must keep the plant open long enough for the buyout to occur (six months or more), and often must help persuade creditors the deal can be made. Although a few buyouts have been accomplished without this help, its absence makes the odds very long.

Another common factor in many of these buyouts is that the companies were at least marginally profitable—but not quite profitable enough to justify continued investment by outsiders. Employees, however, viewed any profits as enough, as long as their jobs were maintained. If the companies were not profitable, there were clear problems an employee buyout could solve. At Hyatt, for instance, low productivity was a key problem. Once the employees owned the company, productivity increased dramatically, largely because of work-rule changes the union would not allow when GM owned the plant.

In other words, if a company can clearly demonstrate that as an employee-owned firm it can make a profit and repay its loans, it can generally secure private financing, albeit often with the help of the previous owner.

TRADING CONCESSIONS FOR OWNERSHIP

Another increasingly common approach is to trade wage concessions for stock ownership. Usually, employees acquire a substantial minority position in the company as a result, but in a number of cases, they have been able to bargain for 51 percent or more.

In the trucking industry, 12 firms, with from 800 to 5,000 employees, have become substantially employee owned in this manner, all within the past two years. Five failed within the first several months, however. Transcon provides a typical illustration.

Transcon is the nation's tenth largest motor freight carrier. During 1981, it had a net loss of $785,000; in 1982 the loss figure soared to $8.8 million. The losses were caused primarily by two factors. In 1981, deregulation of the trucking industry took effect, and 10,000 new firms entered the field. Most of these companies were small and non-union, and paid their employees considerably less than the unionized Transcon did. This new competition came at the worst time, for 1981-1983 saw the deepest recession since the depression. Transcon needed to lower its labor costs (its largest expense) and raise new capital for more modern equipment. Banks, of course, were reluctant about investing in a firm in such financial difficulty.

In mid-August 1983, Transcon's management decided that an ESOP was the only answer. They worked out a cooperative agreement with the Teamsters to ask the company's 3,700 employees to take a wage cut of 12 percent for five years. The cut would be outside the Teamsters' contract, and would be entirely voluntary. Employees who accepted the cut would receive stock in return (but the value of the stock would be less than the value of the concessions); 88 percent agreed. The result was that employees will own 49 percent of the company through the ESOP. Since employees already owned 8 percent, they will eventually own 57 percent of the voting stock of the company. They can currently elect two members to the thirteen-person board (salaried personnel have an additional seat reserved), but as their voting power increases, they will be able to elect more board members.

The effect on the company has been dramatic. After acceptance of the plan, the value of the company's stock, which is publicly traded, went up over 400 percent. The company had a profitable first quarter in 1983, just before the plan, and now seems back on the road to sustained profitability.

The stock-for-wages pattern, common in trucking and in the airline industry, is already spreading to other areas as well. Although employees will not normally receive majority ownership, it seems likely that this will become a growth factor in buyouts and employee ownership generally. It is one case in which employees are risking relatively little, because before the stock-for-wages swap approach became popular, employees were normally faced with a demand either to take concessions or face the loss of their jobs and were offered no stock in return.

FEDERAL ASSISTANCE FOR BUYOUTS

To our knowledge, sixteen employee buyouts have received grants, loans, or loan guarantees from the federal government. Nine companies received help from the Economic Development Administration (EDA), four from the Department of Housing and Urban Development's Urban Development Action Grant Program, one from the Farmers Home Administration, and two from the Small Business Administration. Typically, management or an employee group approaches an agency, often with the assistance of a local political leader or member of Congress, and seeks a federal grant or loan. In most cases, the effort is futile. The company is often closing too soon to make it possible to go through the months of paperwork needed to receive government assistance. Even if there is time, the business prospects are often too discouraging. Finally, if time and feasibility are present, federal money often is not. Especially in the past two years, the agencies that have been involved in buyouts have seen their budgets shrink significantly. If all the federal agency budgets available for worker ownership were added together, it is doubtful that more than $20 million could be allocated in any one year—about 7 percent of the Weirton Steel price tag, 40 percent of Hyatt's, and 100 percent of a medium-sized manufacturer.

To the extent that it is available, federal aid comes in three forms. EDA's and HUD's UDAG program generally makes grants to local government or economic development organizations, which then loan these funds at low rates of interest to a newly formed employee-owned firm. EDA, for instance, provided a substantial proportion of the purchase price of South Bend Lathe Company and Okonite, two of the early employee-buyouts, by making grants to local development groups. HUD recently assisted Weirton Steel in this fashion, with a $10 million grant to the city, which was then loaned to the company.

Outright grants are less common. The case of Rath is a rare exception. Loan guarantees are much more common. The Economic Development Administration, for instance, guaranteed a loan for the buyout of Jeanette Sheet Glass and the SBA has guaranteed loans to two small buyouts (a poultry firm that subsequently closed and the O&O markets in Philadelphia). Regardless of the form of the aid, however, it seems unlikely that the federal government will play a major role in future buyouts.

STATE ASSISTANCE FOR WORKER BUYOUTS

As the federal role is declining, the state role is increasing. California, New York, Illinois, Michigan, and Pennsylvania have all recently enacted laws to help finance buyouts, usually through the sale of state bond issues. Funds raised from these issues are loaned at somewhat below market rates to buyout efforts. Generally, the state cannot finance more than 25-40 percent of the total package. The state laws also generally require that employees have full voting rights on their stock. More states are considering legislation, although others are using existing laws to help buyouts. Michigan also has in place a program to provide technical assistance for worker buyouts, and California, New York, and Illinois are considering such programs as part of their own existing efforts. State financing programs are unlikely to put up more than several million dollars per year for buyouts, but the cumulative effect of such programs in ten to twenty states could be far larger than all the existing federal programs. The states, therefore, are likely to become the major source of government assistance in the future.

State involvement in a buyout is useful in two ways. First, the lower cost funds are often a key, if minor part, of a financial package. Second, state backing can help "leverage" private money. By providing a loan or loan guarantee, the state, in effect, is reducing the risk that private lenders must take. Because state involvement is relatively new, only a few state initiatives have occurred. Perhaps the first was a loan made by the Pennsylvania Industrial Development Authority to Jeanette Sheet Glass.

National unions have not provided financial assistance for employee ownership, other than a $10,000 grant to employees of Atlas Chain in Pennsylvania from the UAW to help them conduct a feasibility study. Nonetheless, this chapter would not be complete without mention of a local union's efforts in Philadelphia to organize worker buyouts. Although it is a unique case, it creates a model that other unions might conceivably follow.

In November of 1981, A&P closed its Philadelphia stores, putting 2,000 people out of work. The local union (the Food and Commercial Workers) commissioned a study to determine if a buyout of some or all of the stores would be possible. The story becomes complex at this point, but essentially, the result was that A&P agreed to reopen over half its stores if workers would take concessions. In return, the union asked for and received a worker participation program and a bonus fund into which 1 percent of the new stores' gross sales would be placed (a great deal of money in the supermarket industry). Initially, 35 percent of this fund was to be used to set up an investment fund for worker buyouts (that has subsequently been reduced considerably) and 65 percent for bonuses to workers, provided labor costs were kept below a certain level.

At the same time, workers were pursuing the option of buying some of the stores and setting them up as worker cooperatives. Ultimately, more than 600 workers said they were willing to pledge $5,000 each to help buy out stores. As it turned out, however, the union decided to bid on only two stores, with an option to bid on others. About 60 workers were involved initially. Their own $5,000 apiece, plus a loan guarantee from the SBA and other local private sources, funded the two new stores, dubbed "O & O." After their first year, the stores were exceeding expectations substantially, and efforts are now under way to use the investment fund and other funds to open a third store, as well as a number of smaller stores that would be a combination of small supermarket and a convenience store.

The local union was critical in organizing this effort, and with the help of local consultants, initiated the investment fund idea. Although the fund is smaller than originally planned, it still is accumulating a substantial amount and, as the O&O stores grow, it is certainly possible that a significant network and funding capacity can be created (New York Times, 1983; In These Times, 1982).

Similar nonunion efforts are under way in North Carolina and Boston, where local nonprofit groups have tapped foundation and private funds to establish funds for cooperative development. Although only some of these monies will be used for buyouts, the efforts being made to create a private source of capital devoted to employee ownership are intriguing.

CONCLUSIONS

The experience with financing employee buyouts suggests several important lessons. First, most employee buyouts have succeeded. This success, and the publicity it has received, has made it considerably easier for buyouts to attract private funding. Nonetheless, it is clear that such funding will be available only if banks and other creditors are assured that they can recoup their loans even if the company should fail. That generally means the company must have valuable assets as collateral or a loan guarantee or subsidy (for example, a second party that will take a subordinated position—that is, will agree to let the bank have first access to any assets). If these conditions are not met, the company must have very strong indicators of an ability to turn earnings around. It also seems probable that private funding will require something less than complete or immediate control of a company by the workers, although private capital has certainly come a long way on this issue in recent years. Most major lenders seem willing to have some employee representation on boards in buyout cases; in at least one case (Atlas Chain) they agreed to have a majority of the board elected by workers. Just as it took time and experience for private sources of capital to become familiar with the idea of buyouts, it will take time for them to become comfortable with greater employee-control (assuming that that control does lead to economic success or at least stability).

The fact that most employee buyouts have been largely or wholly privately financed reflects the fact that most of the buyouts have been of companies that were either marginally profitable when the employees bought them (but not profitable enough to justify the continued investment of their previous owner) or the companies possessed some clear factor that, when removed or added, would restore profitability (such as low productivity or high labor costs). Buyouts of companies in less certain circumstances will either require government assistance or will be the result of stock-for-wages concessions in which the concessions are matched by enough shares to constitute a majority of the company's stock.

Government assistance is likely to shift to the state level, where it will generally form a minor part of the financing package. Although this will lower the threshold for participating private capital, it will still require that buyouts be able to demonstrate very strong economic feasibility. At the same time, given the way state laws are being written, the buyouts are much more likely to require full voting rights for employees over their stock (although there may well be transition periods before the employees can elect a majority of the board).

Regardless of the source of capital, employees will be faced with several obstacles in putting a buyout together. These include adequate time to arrange the buyout before the company closes and customers and markets are lost (this usually means a range of from six months to two years), well-qualified management willing to work in a risky situation *and* one in which employees generally have at least some input into management, access to capital at rates that can be repaid, and initial funds for a feasibility study (although some states may help with this). Moreover, the current owner must be willing to cooperate. Some plants are closed and their production shifted elsewhere; some are closed in an effort to reduce capacity; some are closed by owners who need tax write-offs. These and other considerations may make a current owner unwilling to negotiate, yet without the owner's cooperation, buyouts are extremely difficult to organize.

Given all these obstacles, along with the basic question of the economic feasibility of the company, it is not surprising that there have not been more employee buyouts. At the National Center, for instance, we would estimate that no more than ten to fifteen buyouts are likely to occur in any one year, although as many as one hundred may be tried. Buyouts are dramatic and important instances of employee ownership, but these numbers should place their significance to the overall employee-ownership movement in perspective. The ultimate success of that movement will depend on convincing new and ongoing successful businesses to incorporate employee ownership. Fortunately, considerable progress is being made in that direction.

NOTE

1. The best chronology of the buyout is to be found in *Metal Producing* (1983).

REFERENCES

ELLERMAN, D. (1980) "What is a worker's cooperative?" Industrial Cooperative Association, Somerville, MA.
HAMMER, T. and R. STERN (1980) "Employee ownership: implications for the organizational distribution of power." Academy of Management Journal (March).
In These Times (1982) "A big step towards worker ownership." (June 2-15): 2.
McCLAUGHRY, J. (1974) "Employee ownership: a new way to run a railroad." Business and Society Review (Spring).
Metal Producing (1983) "Weirton: the acid test." (April).
MURNANE, T. (1984) "Another ESOP fable." California Business Magazine (March): 55-60.
National Center for Employee Ownership (1983a) An Employee Buyout Handbook. Arlington, VA: Author.
———(1983b) An Employee Ownership Reader. Arlington, VA: Author.
———(1983c) A Legislative Guide to Employee Ownership. Arlington, VA: Author.
New York Times (1983) "A&P's worker-managers." (May 21): B1, 3.
SIMMONS, J. and W. MARES (1983) Working Together. New York: Knopf.
WHYTE, W. F., T. H. HAMMER, C. B. MEEK, R. NELSON, and R. STERN (1983) Worker Ownership and Participation. Ithaca: Cornell University, New York State School of Industrial and Labor Relations.

15

LABOR POLICY AND
THE CHANGING ROLE OF GOVERNMENT

Joseph Blasi

The signs are now clear that some important changes in the relation between labor and management have begun in the United States. These changes might have wide-ranging and unexpected impacts on what the labor economy will look like in the next few decades, the future of labor relations, and the role of government in labor law and industrial relations. This chapter will consider the role of worker involvement in ownership and participative management. There are two arguments. The first relates to the possible growth of employee ownership. The second relates to the impact of employee ownership and participation on labor relations, and the role of government, unions, and management. It should be clear from the outset that the concerns we raise about the structures of employee ownership and participation and their legal and ideological basis stand as problems whether employee ownership grows a lot or a little. This chapter is an attempt to categorize the myriad current events in this area and raise some broad issues that might have significant policy implications. It is certainly an attempt, albeit risky, to transcend the more detailed discussions of what is being done and how it is being done in the area of employee ownership, and search for the deeper meaning of trends that are only beginning to be discernible. It is also an attempt to transcend the current limitations of research in this area, which almost exclusively consists of case studies focusing on the internal organizational variables of the firms involved.

Companies that are wholly or partly employee-owned might represent a major part of the labor force and have more workers than those currently organized by trade unions. If this occurs it will present yet another new reality to the trade union movement. How they react will determine both their own role in the new labor economy and their ability to continue to represent workers' interests in light of changing conditions. Whatever the extent of the development, the traditional roles of labor and management will be reversed. Workers will own and managers and entrepreneurs will work with or for workers in increasingly cooperative and innovative companies. A new form of business or labor organization will emerge, the employee-owned firm, a *labor-business organization.*

Government's involvement in labor law and labor policy through the National Labor Relations Act (NLRA) has protected workers' rights and management rights given the *traditional* roles of both groups. Federal law governing employee-owned firms was written when employee ownership was often used as an accounting device to introduce small amounts of stock ownership into companies. Current federal law does not *mandate* voting rights for workers who own stock. There is increasing evidence that workers' rights can be threatened and undermined in employee-owned firms without the proper safeguards. In short, I will argue that current federal employee-ownership law and labor law cannot sufficiently safeguard workers' rights. And, given the extent to which employee ownership and labor-management cooperation might grow, there is reason to believe that this growth may in effect hasten the deregulation of labor policy unless it receives sufficient attention. Trade unions are used to protecting their members' interests as adversaries, but they are unsure what to do in a company they own.

It is yet unclear whether trade unions will organize the new sector of worker-owned firms or not, or whether or not the nonunionized labor-business organization, the employee-owned company, will replace trade unions as the dominant form of associated labor organization. Given the continuing decline in union membership and organized labor as a percentage of the private-sector work force, this cannot be immediately dismissed as a possibility.

Management and entrepreneurs will have to learn to involve workers in decision making to tap the tremendous potential to increase productivity through labor-management cooperation, and increasingly fulfill their role as "business associates" of workers rather than as bosses, and workers will increasingly realize that a narrow definition of employee ownership as simple equity participation has the potential to significantly undermine their old rights as workers and their new rights as entrepreneurs. And if workers take their rights seriously in an employee-owned company, they will ultimately be responsible to demonstrate their utility as entrepreneurs by improving the business through concrete programs of labor-management cooperation that can promise definite results. If this does not take place, then a narrow vision of worker ownership and participation will become dominant that says it is exclusively either an alternative benefit plan or an alternative investment strategy involving workers. This chapter will assess the implications of both a broad and a narrow definition of employee ownership and participation.

THE CONTEXT

In 1974, federal tax law introduced the ESOP, the Employee Stock Ownership Plan, as a new institutional form in this country. Legally, ESOPs are stock bonus plans and are defined as deferred employee-compensation plans under the Internal Revenue Code. The meaning of the "ownership" the worker receives will depend on the following variable characteristics of ESOPs:

How an employee is defined. Usually the criterion used is that employees must have completed one year or 1000 hours of service, but ESOPs can exclude hourly employee or union members and set minimum ages, although the union must have the right to bargain about involvement in the ESOP in unionized companies. One of the problems of ESOP law is that, although it uses the notion that the employee ownership must be nondiscriminatory, in fact, many discriminations are allowed and the principles on which the allowed and the disallowed discriminations are based must be more clearly defined.

Presence of voting rights. Different types of ESOPs can have voting or nonvoting stock or combinations of the two. In all cases the ESOP trustees vote the stock, so the issue is whether these voting rights pass through (the trust or the ESOP trustee) to the employees. One usual practice is not to allow voting stock in leveraged ESOPs until the loans to the commercial lender used for the transaction are paid off.

How voting is structured. ESOPs can have passed-through voting rights structured according to one vote per worker no matter how many shares he or she owns, essentially like a producer cooperative (Ellerman, 1984), or according to the number of shares an employee owns, the predominant approach.

How shares are allocated. Shares can be allocated among the employees according to salary and (or) tenure, hours worked, or equally. The law says that in leveraged ESOPs the shares can be allocated only as the loan is repaid and this must be viewed as a further stipulation of ownership in that type of company.

Who selects the ESOP trustees. The trustees can be selected by management, the board of directors, the commercial lenders involved in setting up the firm (if it was a leveraged buyout by the employees), the union, the law firm, or a combination of several actors. They can be part of the company or outside of it and beholden to different interest groups. The trustees have special fiduciary responsibilities defined by federal law and occasions can arise when these responsibilities conflict with directions they get from employee-stockholders on how to vote shares of stock. In such a case they can either overrule the workers or step down.

The vesting schedule of the stock. "Vesting" refers to the nonforfeitable right of the worker to his or her trust assets. Each ESOP sets up a schedule that determines when a worker's stock account is vested with that worker. The law requires full vesting within ten or fifteen years depending on various considerations, although it is a common practice to vest stock at about 10 percent a year for ten years. By not vesting employees for as long a period as possible, the "ownership" of the stock can be substantially out of the control of the employees for a very long time.

The manner of distribution of the stock. Three aspects are important here: when employees receive their capital, whether they can sell their shares, and how they are valued at the time. Distribution refers to when employees receive the market value of the vested portion of their account. It usually happens after any permanent break in employment. In some companies it may be when the employee reaches 55 or 65 years of age. This is true even if the worker leaves the firm at a very young age and has stock less than 50 percent vested. Usually, a worker cannot borrow against the vested portion. In closely held companies the employees usually have a put option requiring the firm to repurchase the employee's shares at fair market or appraised value within the first year of termination of employment or retirement. Some companies pay the value of the shares in a lump sum and others may pay this out over a number of years, whereas in publicly traded firms the workers can simply sell the shares on the open market. In the closely held firm an independent stock evaluator establishes an appraised value, whereas in a publicly traded firm the fair market value is used. The ownership rights of the individual employee can be curtailed by delaying distribution as much as possible. The ownership rights of the employees as a group can be curtailed by arranging that the firm go public as the stock touches the workers' hands (that is, when it is distributed to them), so that, in effect, employee ownership is just a temporary financing tool until such time as "someone" determines to remove the employees as investors.

Percentage of ownership. The percentage of ownership may be very large or very small. The percentage of ownership in legal terms means the amount of stock in the ESOP trust whether or not it has been allocated to employees or fully vested. It should be clear that per se the percentage of ownership as defined does not indicate anything about the level of employee control of the company until it is examined in tandem with other ESOP characteristics.

Who chooses the company board of directors. This is not typically considered an aspect of the ESOP but since the board of directors in any ESOP may have an important role in determining other ownership features of the ESOP, it is significant. The board of directors determines, for example, the size of the company contribution to the ESOP. The minimum tax deductible contribution is 15 percent of the covered employees' payroll although under certain conditions it may be greater. Through this it will determine how to evaluate various financial aspects of the ownership offer to employees. This decision may be based solely on the tax requirements of the company, or may include various motives for sharing or not sharing the profitability of the company with the employees. If the board of directors has the right to appoint the ESOP trustees or is itself self-perpetuating, it may be able to affect the ownership rights of the employees substantially over time. It can plan to take the company public under certain circumstances without consulting the employees.

These nine factors can be enumerated and evaluated for every ESOP. The sum total of their effect determines what the real motives were for setting up the ESOP, who has power, what the actual definition of ownership for that ESOP is, given the fact that employee ownership is a vague flexible term in actuality. Obviously, one version of the most nondemocratic ESOP would exclude significant groups of employees, have nonvoting stock, be based on allocations to more highly paid and senior employees, let management elect the trustees, establish a long vesting period, distribute shares on retirement, have a low percentage of "employee ownership," and have a board not chosen or influenced by employees. It is, however, not easy to evaluate ESOPs because, as this structural analysis will prove, one can combine a large number of the more pro-employee democratic characteristics of an ESOP with simply one or two nondemocratic characteristics and end up with a largely undemocratic ESOP. For example, as long as workers do not have voting rights or do not elect the board of directors, all other ESOP characteristics can be democratic, but the workers will have little influence. Also, the company can be structured to be completely democratic but have a provision that it will go public on a certain date. Or the ESOP can be designed differently for management and worker groups, with management groups having voting stock purchased at lower prices and control of the board and trustees, and workers having few of these ownership rights. These hidden combinations are often referred to as "time bombs" in an ESOP, and they represent an important aspect of the labor-relations impact of employee ownership.

Therefore, it is meaningless to talk about "employee ownership" unless one simply means the prevalence of the narrowest kind of legal ownership. It must be noted that producer cooperatives are employee owned as are joint-stock companies not set up through ESOPs. ESOPs have become the prevalent form of employee ownership because they are favored by tax legislation. Every ESOP includes a philosophy of "employee ownership" embedded in its structure that defines the motives for setting it up, the structure of power between management and labor that is envisioned, and the ownership rights of the participants. David Ellerman (1984) has defined the democratic ESOP and suggested a legal and philosophical basis for this form. His work represents the most serious attempt to date to analyze the ESOP in a fundamental way. We will now attempt to estimate trends in employee ownership and then analyze the meaning of these trends.

ESTIMATING TRENDS IN EMPLOYEE OWNERSHIP

It is not necessary to know the conceivable trends in employee ownership in order to analyze the possible labor relations problems in specific firms. But, the impact of employee ownership on the labor economy, especially organized labor, and the expectations workers in employee-owned companies can have about the nature of their labor relations and their rights, requires some attempt to project the importance of this phenomenon.

This study is based on several sources of data. First, a three-year
review of a national newspaper clipping service on developments in
employee ownership has allowed us to monitor developments and track
trends, and estimate the direction of change. Special emphasis has been
given to monitoring significant business publications such as the *Wall
Street Journal, Business Week, Forbes, Fortune,* the *New York Times,*
and *Inc.* magazine for small businesses. Second, interviews with
professionals who are involved in establishing employee-owned companies were conducted to check the validity of these trends. These are
technical assistance organizations, law firms, investment banks, and
trade union officials. Third, we have projected the effect recent federal
legislation giving major tax incentives for establishing employee-owned
companies will have on the growth of this phenomenon. These were
contained in the Deficit Reduction Act of 1984 signed into law by
President Ronald Reagan in July 1984. Fourth, we have taken estimates
of the growth of employee ownership in the past five years, and
telescoped them into the future, presuming basically the same rate of
growth. Currently, this is the best way to gauge the development of
employee ownership because the relevant surveys are based on small
samples of companies and were mainly done in order to assess
organizational features or economic performance of the firms (Tannenbaum and Conte, 1977; Marsh and McAllister, 1981; ESOP Association
of America, 1982, 1983; Rosen and Klein, 1983; Klein, 1984; Tannenbaum et al., 1984). Steven Bloom (1984) has done the most comprehensive study of a large sample of ESOP firms to date, but it includes only
1,918 firms and 5.2 million workers and does not provide data on most
of the nine characteristics discussed above. Thus existing surveys either
discuss some of the key characteristics of ESOPs for small samples or
are made up of larger samples for which detailed organizational data are
unavailable. These studies do not attempt to estimate interest in
employee ownership or potential growth. Within the next two years
several surveys will be completed that will allow a more detailed and
careful estimate of employee ownership. Nevertheless, given current
information on the growth of employee ownership, clear evidence of
patterns that raise serious issues, and the public policy implications of
these issues, it is important to examine employee ownership as carefully
as possible at this point in time.

We estimate that within thirty to fifty years there will be more
workers in companies that are more than 20 percent worker owned than
in firms that are currently unionized. This is a conservative estimate and
there is reason to believe that if employee ownership continued to grow
simply at the rate it has over the past decade that the workers of

employee-owned firms of this type may constitute an even larger part of the labor force. Today the organized work force makes up 17 percent. If it is true, as some observers (notably, James Medoff of Harvard University) project, that this could fall to 10 percent within thirty years if current trends continue, then our estimate may be somewhat conservative (Boston Globe, December 8, 1984: 11).

Since current data do not allow exactness, the important observation is that employee ownership will be a major factor in entrepreneurial organization in our society and not merely a minor experiment. The employee-owned labor-business organization will emerge as a significant form of labor organization. At present, we are making no claim regarding how many of these workers will be members of traditional unions. If trade unions decide to organize the small business sector where most of the employee-owned firms are and conceivably will be concentrated, then the growth of both movements may happen largely in tandem. What is important is that roles in this sector will shift from workers being simply employed to owning and participating in management and owners, entrepreneurs, and managers will increasingly be working with or for groups of worker owners, creating a de facto business partnership between them.

Given the ongoing decline of the private-sector trade union movement, the mounting erosion of master wage agreements and the nation's labor policy and union's political power, along with strident anti-union activity combined with more appealing human resource management practices, the importance of this trend will be more pronounced (see Bloom and Bloom, 1984). The National Center for Employee Ownership estimates that there are 7 million workers in about 7000 Employee Stock Ownership Plans (ESOPs) today. It is estimated that 1 to 1.5 million of these workers are in firms that are more than 20 percent worker-owned and 500,000 workers are in a subgroup of 700-900 firms that are 51 percent or more worker-owned. This represents a 25 percent growth in employee ownership since 1981, the year for which the most recent Internal Revenue Service computer tapes are available, when workers in all such firms already constituted 7 percent of the private-sector labor force.

THE IMPACT ON LABOR POLICY

Perhaps the more important development is the potential for a substantial undoing, undermining, or transforming of labor protections and laws as we currently know and understand them as a result of the clouding and switching in roles of labor and management. It is my argument that the relations between labor and management in a conceivably large sector of the work force will no longer be traditional. Although we do not suggest overlooking the possibility of increased cooperation between labor and management, the possibility that employee ownership will become a new battleground cannot simply be brushed aside. Labor relations cannot and will not be regulated with the same actual effect by the traditional approach of government to labor law, unions to their function, or business in its view of management rights. All these were designed with the traditional worker and the traditional company in mind, not the worker-owner and the manager-worker or entrepreneur-worker. True, there is a potential here of a new system of cooperative business-labor organization with new forms of governance and management in which the rights and obligations of labor and management are not solely oriented toward their meeting in battle, but it would be wishful thinking to project exactly what the organizational life of each firm in this new sector of the economy will be like.

The following observations will illustrate the stress on the system of labor protections, which in light of our projection of the future size of worker ownership, spells a coming crisis in the labor market identity of astounding proportions.

CONSTRAINTS ON THE RIGHT TO VOTE

First, a "free-fall zone" in federal law governing Employee Stock Ownership Plans, the major legal vehicle of employee ownership defined in U.S. tax laws in the Employee Retirement Income Security Act (ERISA), is posing, and, given our projection of the future number of worker-owners, will pose a serious threat to worker rights. When workers become owners they receive stock, and with their expanded risk, involvement, and identification with the company they have a potential right to vote on the affairs of that company. ESOP law says that in publicly held companies with worker ownership the shares must have voting rights, but in closely held firms, the firms can be structured without workers having any right to vote their stock, even in majority-owned or 100 percent worker-owned companies.

Earlier, the complex of strategies for limiting these rights was analyzed. The *predominance of closely held companies without voting rights* is what makes this a serious issue. The National Center for Employee Ownership estimates that over 95 percent of all majority worker-owned companies are closely held. This is also true of firms 20-50 percent employee owned. Of the estimated 7000 ESOPs, NCEO projects from the Marsh and McAllister data that 85 percent do not give voting rights to the workers. Companies that are more than 51 percent employee owned do not give workers voting rights in 43 percent of the cases, according to an NCEO survey (Rosen and Klein, 1983), and companies that are 20-50 percent employee owned tend to restrict voting rights even more, although sound estimates are not available for this group.

There is reason to believe these estimates are high, because they are based on nonrandom samples that probably overrepresented firms with worker participation, but the trends establish the reason for concern quite clearly. Since closely held firms account for the major growth of worker ownership, the governance aspect of these firms will now be strongly affected by this "free-fall zone" in the federal law. For reasons of space and because existing research has not comprehensively examined the prevalence of various other combinations of ESOP characteristics in limiting worker rights, we are focusing on voting rights. It should be clear that, given ESOP practice, *even those ESOPs with voting rights probably have present other characteristics that nullify their democratic character.*

When a publicly held firm, such as Dan River Textiles, converts to employee ownership, it becomes closely held, and a further strategy for eliminating the necessity of worker voting rights is introduced. There is no reason to believe that new growth in employee-owned firms will substantially alter the governance involvement of workers, because most of the individuals setting up such plans around the country are ideologically committed to a traditional definition of a worker's responsibilities even when he or she becomes the owner. Some groups, such as the Industrial Cooperative Association of Somerville, Massachusetts, and the Philadelphia Association for Cooperative Enterprise have accurately analyzed this problem for years and they specialize in creating democratic ESOPs or worker cooperatives, but the problem cannot be principally explained by the lack of professional consultants to structure these companies in an alternative way. There is no reason to believe that the trend toward restricting workers' voting rights will continue because federal law allows it, and since it coincides with the ideological bent and power position of the major actors initiating ESOP design. The worker-owned trend will de facto be a trend toward reversing shareholder-voting rights when shareholders are workers in about half the cases *by conservative estimates* and, by extension, severely limiting worker rights when they become owners.

When a worker in the United States owns stock in a company where he or she does not work, he or she usually has voting rights. Yet, when an individual owns stock—even all the stock—in a company where he or she works, there is a 50 percent chance that he or she has absolutely no voting influence. This chance is protected by federal law and circled by a variety of other helpful potential aspects of ESOPs. Under ERISA, when trustees of a workers' pension fund invest the money, they must abide by a "prudent man rule" that dictates prudent investments, yet what rules of prudence secured worker-rights in the above cited examples of majority worker-ownership of their own companies? Why must a worker in a closely held firm not have as a basic right the federally regulated right to vote his or her stock that workers in a publicly held worker-owned firm have? It is true that the abandonment of this right is voluntary; the law simply allows voting or nonvoting stock as an option, but case studies indicate that workers have little understanding of ESOP law; many employee purchases of closely held firms or conversions of publicly held firms are management, entrepreneur, investment banker, or lawyer initiated; and usually these parties do not have workers' power or influence in mind. Unions, because of their hostile or ambiguous attitude toward worker ownership, have not provided the information or expertise to examine other options, or have not been interested in helping locals confront this issue (see Whyte and Blasi, 1984).

Workers often accept their new role on a silver platter and little effort is made to examine the implications for the new form of labor-business organization in which they must now function. Many ESOP enthusiasts believe employee ownership should not have voting rights. The ESOP Association of America is currently trying to lobby for legislation that would further restrict voting rights for workers in closely held companies. That such an amendment could be quietly slipped into any tax law seriously affecting the future of American labor relations given the projected growth of employee ownership is alarming. Restriction of workers' rights to vote in these firms in light of how it comes about, the American tradition of democracy and the general predominance of voting stock, the inconsistency with federal law regarding publicly held companies, and the projected growth of employee ownership, dictate that the policy implications of this state of affairs be examined urgently. This is a major issue of workers' rights that should concern American trade unionists.

Why should workers have fewer rights when their roles shift to those of owners? I am arguing that employee ownership could willy-nilly include a political agenda of reducing worker influence the more they become involved in companies. A basic principle in the tradition of labor law in the United States has been government enforcement of democratic procedures in labor organization vis-à-vis management power in business organizations. United States labor law limits management rights. This is one of the fundamental ways the government is involved in business in the country. While employee ownership can lead to more labor-management *cooperation,* the potential for participative management, increased worker commitment, and government encouragement of industrial innovation will be cut short if this ownership is based on labor-management *cooptation.*

Employee ownership law through ERISA makes democratic organization in the new labor-business organization that is emerging, simply optional and often confusing, in effect, deregulating the spirit, if not the letter of our national labor policy. ESOP law's labor policy implications must be addressed and the right to vote all shares in closely held companies and democratically elect the trustees of the Employee Stock Ownership Trust must be mandated by federal laws. Complex combinations that use ESOP characteristics in tandem to nullify democratic procedure should be made illegal except in cases in which allowing nonvoting shares will actually or eventually secure greater worker-rights. These cases are those in which (a) nonvoting shares are necessary in a gradual conversion from a traditionally owned small business to an employee-owned firm in which a small businessperson being bought out over time (before his or her eventual retirement) does not have to compromise control or assets of the firm he or she built until the workers actually own a majority stake, or (b) nonvoting shares are used to facilitate the hybrid ESOP-cooperative organization in which, technically, all shares have no voting rights, but cooperative bylaws in the ESOP allow each worker one vote to instruct the trustees of the Employee Stock Ownership Trust how to vote the shares.

OTHER CONSTRAINTS ON THE LABOR ORGANIZATION

The traditional approach of labor law may need rethinking as a new form of labor-business organization emerges. The National Labor Relations Act (NLRA) protects workers' rights to self-organization, collective bargaining, and democratic procedures to constitute a labor organization, all as government-enforced, *not* self-enforced, rights. Yet it is now evident that from a functional point of view, in which labor is both owner and worker, it might follow, given the tradition of American labor law, that the right of self-organization, collective consultation about company policy, and democratic procedures in how labor organizes itself, becomes a right *internal* to the firm and not simply of the union outside the firm. Why should labor's ownership reduce its power to represent itself, or, at least, why should a reduction of labor's power of self-representation be sanctioned by legal forms of employee ownership? Employee ownership is not simply workers deciding to go into business; it involves a more complex relationship of existing workers to groups of managers and entrepreneurs, and often, bankers and commercial lenders.

This view is bound to upset everyone with traditional institutional biases. The trends, however, are clear, and the biases are increasingly irrelevant and deviant themselves. According to a recent New York Stock Exchange study, *People and Productivity* (1982), which was the first broad-based survey of human resource programs emphasizing increased labor-management in decision making or gain sharing, 1 in 7 companies with more than 100 employees covering half of all corporate employees have some kind of program involving 13 million workers. The driving forces are increasing competitiveness, improving productivity, and cutting costs. Companies report they are undergoing a change in management philosophy and that their efforts are actually increasing productivity, morale, service and product quality and decreasing employee turnover, absenteeism, lateness, and grievances.

What is important, for the purposes of this chapter, is that for the first time ever a major national survey has actually linked the growing phenomenon of labor-management cooperation with the phenomenon of gain sharing. The study found that 15 percent of all U.S. companies with more than 500 workers have some kind of gain-sharing plan, including employee ownership. The major common characteristic of these companies is that they maintain an open and participative atmosphere. For example, companies with stock purchase plans of all types including ESOPs are almost four times more likely to have quality circles than firms of comparable size in the United States. Nonmanagement employees participate in decision making in 63 percent of companies with group productivity programs, 70 percent of companies with profit sharing, and 82 percent of companies with stock purchase plans. Although this study has several limitations because of the broad way it measures work innovations, and although it is far more optimistic than I am about the speed, depth, and real participative nature of this phenomenon, the direction of the trend is unmistakable: extensive experimentation in changing traditional roles is taking place within U.S.

work organizations. If anything, the confusion between the nondemo-cratic trends in worker ownership and the participatory trends in labor-management cooperation may indicate a period of transition. Perhaps no top-level American company illustrates this thrust as much as Eastern Airlines, with its combined employee ownership and employee involvement and productivity task-force program. This case is now the subject of a year-long research program coordinated by the Harvard Study Group on Worker Ownership and Participation in Business (Blasi et al., 1984). As a largely unionized company, Eastern has at least succeeded structurally in protecting labor's right to self-organization, consultation with management, and democratic proce-dures, when labor entered the company to share the role of entrepreneur. A lot about this problem will be learned from that case.

It is unacceptable that when workers are not owners NLRA limits management rights clearly in unionized firms and companies in the process of deciding on unionization, yet when workers are owners we can have the complete elimination of workers' voting rights or substantial weakening of labor's capacity for self-organization, within the new labor-business organization. Strangely enough, by forcing the adversarial union outside the new firm—if it exists—to be labor's only source of influence, designers of nondemocratic, employee-owned firms set up a situation that will create conflict, even strikes, as was the case at South Bend Lathe and the Okonite Company in New Jersey. In excellent contrast to the responsible character of the Eastern Airlines case, the South Bend Lathe case illustrates that this weakening is not solely a problem of voting rights.

As part of a Cornell study of worker buyouts, including scholars from several universities in which William Foote Whyte and I are involved, economist Charles Craypo of Notre Dame University has been examin-ing South Bend Lathe. This is a 100 percent worker-owned firm with pass-through of voting rights, although management persuaded workers to agree to a management-appointed board and a group of Employee Stock Ownership Trust trustees that are self-perpetuating. Workers have had trouble obtaining the most elementary information from the management of their company. Management persuaded them originally to give up their independently secured pension plan and view the company stock as protecting their future as a strategy to save their jobs. What emerges is not exactly job preservation. It is now alleged that management created a South Korean joint venture through unclear institutional arrangements and transferred many manufacturing opera-tions to that country, thereby reducing the work force from about 500 to about 150. Informed observers say that this transfer of work stands no chance of making a case before the National Labor Relations Board.

ERISA and ESOPs are supposed to benefit workers exclusively. If managers participate in the plan they are also "workers," but it is folly to think that the method of choosing board members, ESOP trustees, and making major business decisions about the company should be totally insulated from workers in a completely 100 percent worker-owned company. In this case, employee ownership has not given workers any more say, and the protections of the NLRA are separate from their employee-owner status. In fact, according to Craypo, the workers' local has also had a very rocky relationship with the United Steelworkers which, in addition to some ongoing tension between the two, is now forcing the local to repay dues it could not afford to pay during a 1978 strike, an unusual demand to make of a local by a national union, according to Craypo. In many cases, when locals are caught in the "free-fall zone" between employee ownership and federal law, the union does not know how or does not want to help itself.

Must NLRA be revised when the traditional roles of labor and management shift as radically as they do in many cases of employee ownership? Posing this question does not mean that all employee-owned firms, ESOPs, or worker coops are now seriously threatening workers' rights, or that all unions are uninformed or uninterested in the face of this reality. (In fact, as we have pointed out, the United Steelworkers is undergoing a substantial shift toward using employee ownership as part of its own strategy.) Neither does it mean that employee ownership law had or has a clear antilabor intent. When Senator Russell Long, the key drafter of the ESOP provisions of ERISA in the early seventies spoke at Harvard University (Long, 1982), he endorsed substantial worker control in worker-owned companies and spoke about his hopes for increased labor-management cooperation. At the time ERISA was drafted, no one could have predicted that ESOPs would emerge as far more than tax-incentive based, minor benefit plans and accounting devices and become part of an emerging sector of labor-business organization the long-term implications of which we are just beginning to understand. Changes in the law and in practice may now be required.

The options are as follows: (a) to amend ERISA's section on Employee Stock Ownership Plans to provide some basic protections for workers' rights; (b) to redefine the National Labor Relations Act to provide for these basic rights; or (c) to acquiesce to a major deregulation of American labor law and policy, and a major, new, and challenging arena for labor (both organized and not), management, and government.

The institutions that are threatened by this are labor, government, and management. Unions will no longer be able to protect workers' rights by using the adversarial methods of outsiders. The adversarial role of the union as commonly understood is much less useful to represent labor's interests if labor owns stock and has no right to vote and influence decisions, is denied the right to choose the board of directors of the firm or even the trustees of the ESOP, or receive ongoing information about the firm. New wine in old wine skins will not work. Unions may need a new role in thinking about employee ownership. The emphasis needs to shift to educating workers and even organizing as yet unorganized workers who are involved (or about to become involved) in such firms using a fresh, novel approach to labor-management cooperation. Unions should be willing to try new things, but must remember that the new experiments in work organization may not be allowed to threaten labor's interests.

THE ROLE OF GOVERNMENT

The government must decide whether its labor policy regarding ERISA and the NLRA requires change in light of the transformations in the traditional roles of labor and management taking place in the labor economy. Should ERISA be updated to confront the reality that employee ownership is now becoming a new organizational form and not a minor, unimportant alternative benefit plan in a traditional company? Is NLRA capable or willing to confront the possibilities that aspects of labor policy will be deregulated as a result of the "free-fall zone" in federal law on employee ownership, and NLRA's inability to regulate labor and management rights when their roles and obligations change? (It is unclear yet whether the problem simply requires substantive attention from NLRA, administrative rulings on test cases, a policy regarding the subject on the part of the Administration or the NLRB, or changes in the National Labor Relations Act.) Surely, the actions government will take depends on whether organized labor formulates a policy on the problem, and how ongoing surveys and empirical research completes our understanding of the issues in question. What cannot be avoided is the possibility that the changes in workers' rights possible in employee ownership, in the name of their interests, represents a resuscitation of the old yellow dog contract. (The yellow dog contract was a contract workers said only a "yellow dog" would sign because it involved workers' agreement not to self-organize and self-govern in their own interests.) Limitations in federal law against passing-through voting rights is a form of federally sanctioned injunction against labor organization and an extension of the union as a conspiracy notion. Allowing financial manipulations of various types without sufficient labor input is simply allowing unfair labor practices of a new and more specious nature. Why should it be an unfair labor practice to close a plant to get rid of a union when workers do not own the plant, but legal to move most of the production out of a plant that workers own and supposedly govern?

Finally, if the government decided that changes in the more cooperative direction in American labor-relations may represent a useful innovation in reviving productivity, competitiveness, and innovation in the economy, then it must squarely face the impact its own legislation on employee ownership and participation is having. In effect, although this legislation is supporting some experiments in labor-management cooperation, in terms of the problems we have outlined, it is encouraging labor-management conflict and a return to the old style of labor relations.

THE ROLE OF MANAGEMENT AND ENTREPRENEURS

Managers, entrepreneurs, and their representatives, who are lawyers, investment bankers, accountants, and organizational consultants, often see employee ownership as simply a financial tool, a strategic device that is, as the ESOP Association of America's newsletter frequently says, "installed" in a company. This view of employee ownership depends on the original motives of the persons who designed the plan, but it should now be clear from my analysis, that the motives of the designers is not a sufficient criterion on which to judge whether or not an employee ownership plan is legally correct, ethically right, and organizationally sound in terms of encouraging a new type of labor-management cooperation. If labor policy and labor law are completely deregulated, then ESOP designers can speak of "installing" ESOP plans in companies as if workers are simply passive participants.

Surely, the trends in employee ownership have an impact on this argument. Employee ownership plans that control less than 20 percent of the company's stock perhaps should be viewed simply as minor benefit plans, and subject to different rules and expectations. Certainly, it would seem that there are discernible trends toward a much greater proportion of sizable (20-50 percent), majority, and wholly employee-owned companies. The direction of federal legislation itself may hasten this trend, if, as expected, the PAYSOP version of the ESOP that led to many minor employee ownership plans, expires in 1987 (Bloom, 1984). Nevertheless, given sizable, majority, or complete employee-ownership of a company, serious questions of legality and ethical principle exist regarding whether or not such firms can appropriately have little or no worker autonomy and regarding the protection of management's absolute right to manage. What behavior on the part of managers, entrepreneurs, or their representatives with respect to these firms is or should be legal or illegal?

Management and entrepreneurs also, like unions, do not simply have negative reasons to examine their own and the worker role in employee ownership. Unions' negative reason is a fear they will lose power in American society, and management's negative reason is a fear its absolute right to manage will be eroded. The positive reason for management and entrepreneurs, as the New York Stock Exchange (1982) study illustrates, is to devise a way to combine smart, well-trained, committed, efficient, flexible workers and managers in a firm in which their human resources are being used more fully and more competitively. Beyond the formal, legal issues of worker involvement in employee ownership is a new world of informal structures to increase labor-management cooperation. Progress in these areas will be the real substance of any definite change that occurs in the ways companies actually operate and innovate.

There are two forms of innovation possible. First, increased economic performance can be the result of the financial aspects of employee ownership, namely, cheaper capital because of tax incentives, reduced or flexible labor costs in competitive markets in which workers and managers choose to cut labor costs or defer remuneration into invested equity, a combination of a base wage with flexible remuneration based on productivity increases, or the ability of workers to act as bankers by when they reduce the cost of capital to their firm through a leveraged employee buyout. If one speaks of the relationship between employee ownership and increased economic performance as being proven and automatic, this is the only way in which the relationship can be said to be fairly direct. It should be clear that this category of innovation is financial and not based on a vague sense of "working harder and smarter" that employee ownership enthusiasts often ascribe uncritically to such companies.

Second, labor and management can custom-design informal procedures for each firm over a broad spectrum that *can* be the source of specific ideas to work harder, work smarter, or improve management, the production process, the product, the market of the firm, or the quality of working life. There is no insurance that such innovations can be translated simply or immediately into improved productivity or economic performance, but numerous case studies and the New York Stock Exchange survey would suggest there is tremendous potential here.

These procedures, in increasing intensity, are as follows: suggestion boxes; open-door policies; increased flow of company information to workers of varying scopes, dealing with different levels of the firm, and inviting different amounts of worker input or decision; worker education to help them understand and use company information more effectively; training of managers in an interactive style that seeks worker suggestions and collaborates with them; establishment of "lead workers" who replace front-line supervisors and reduce supervisory costs; redefinement of the structure of jobs of management and workers; elimination of inefficient work rules; quality circles in the production-related departments; quality circles in all departments of the company; a quality circle coordinating body to coordinate innovation throughout the whole firm; autonomous work teams; companywide task forces set up to solve specific problems and strategic issues; labor-management committees that work on a broader set of issues than quality circles in production-related departments; labor-management committees throughout the firm; a labor-management committee to coordinate innovation; education and training programs and an internal consulting group to teach workers and managers the skills to make these structures work according to the experience of successful companies; workers on the board of directors; experiments with worker-selected managers; attempts to increase the common identification between and among workers and managers by expanding social interaction beyond work activity (for example, day care or educational, recreational, social, or consumer activities in, around, or organized by the firm).

Depending on the number of workers involved, the degree of training and education invested, the quality of management, organization, and evaluation, the range of issues chosen, and the way labor and management originally collaborate in setting up and monitoring the program, there is reason to believe that gradual evolution along this spectrum can bring substantial productivity gains assuming the firm has a strong strategic plan with a good product, a sound management, and a solid market and financial position. If this is not the case, then the first goal of labor-management cooperation must be to design such a plan or it will simply be a story of substantial progress taking place in the context of substantial failure.

REFERENCES

BLASI, J., C. MEEK, and B. SMABY (1984) The Eastern Airlines Study. Cambridge, MA: Harvard University.
BLOOM, D. and S. BLOOM (1984) Institutional Change in the U.S. Labor Market: Some Trends. Cambridge, MA: Harvard University.
BLOOM, S. (1984) The Economics of Employee Ownership. Cambridge, MA: Harvard University.
ELLERMAN, D. (1984) ESOPs and Co-Ops: Worker Capitalism and Worker Democracy. Somerville, MA: Industrial Cooperative Association.
ESOP Association of America (1983) The ESOP Survey. Washington, DC: Author.
——— (1982) The ESOP Study. Washington, DC: Author.
KLEIN, K. (1984) "An analysis of characteristics of employee owned firms." Ph.D. dissertation, University of Texas.
KRUSE, D. (1984) Employee Ownership and Employee Attitudes: Two Case Studies. Norwood, PA: Jerome Weiman.
LONG, R. (1982) Employee Ownership: A Political History. Cambridge, MA: Harvard University.
MARSH, T. R. and D. E. McALLISTER (1981) "ESOPs fables: a survey of companies with employee stock ownership plans." Journal of Corporation Law 6 (Spring).
New York Stock Exchange Office of Economic Research (1982) People and Productivity: A Challenge to Corporate America. New York: Author.
ROSEN, C. and K. KLEIN (1983) "Employee ownership." Monthly Labor Review (August).
TANNENBAUM, A. and M. CONTE (1977) Employee Ownership: Report to the Economic Development Administration. Washington, DC: U.S. Department of Commerce, Economic Development Administration.
TANNENBAUM, A., H. LOHMANN, and G. COOK (1984) Employee Ownership and Technological Change in Companies. Ann Arbor: University of Michigan.
WHYTE, W. F. and J. BLASI (1984) Employee Ownership and the Future of Unions. Cambridge, MA: Harvard University.
———(1982) Recent Developments in Worker Ownership and the Unions. Cambridge, MA: Harvard University.

16

THEORY AND PRACTICE OF
COMMUNITY ECONOMIC
REINDUSTRIALIZATION

Warner Woodworth
Christopher Meek
William Foote Whyte

It is clear that what began in the 1970s as one of the many recessions that have occurred during the past two hundred years of U.S. history has become a period of major social and economic upheaval during the 1980s. The election of Ronald Reagan and the emergence of Reaganomics signaled this change at the outset of the decade. And though an economic recovery of sorts did take place after a deep recession, unemployment has still remained high by previous standards of acceptability. Inflation decreased and in many sectors corporate profits increased, but the tendency has been for these rising profits to serve as resources for financing a flood of merger and acquisition activity that surpasses even that which took place during the late 1960s.

Only a nominal amount of the gains obtained during this period have been reinvested into plants and equipment and therefore there has been substantially less impact upon retaining or creating high-paying industrial jobs than was initially promised. In fact, the increased resource base has to a great extent helped to exacerbate some of the problems that many people had initially hoped it would solve. Indeed, as mergers and acquisitions increase, many conglomerates are simultaneously ravaging regions of the country by closing down plants that represent the lifeblood of communities.

Furthermore, because federal aid has been dramatically decreased at the same time, many communities now find themselves less able to respond to local needs even as the number of unemployed and homeless increases. Thus the "trickle down" that was supposed to result from the implementation of supply-side economics has not trickled down to many and the destruction of America's industrial self-sufficiency looms before us more real than ever before.

In contrast to the hope that passing on more wealth and power to elites would result in the implementation of farsighted investment policies by U.S. corporations, this book attempts to capture some new and alternative approaches that focus specifically upon the social and economic reconstruction of hard-pressed communities through local grass-roots action. The cases described in this volume provide new concepts and models for community planning and problem solving. They suggest tools that have the potential for returning political and economic power to community life. The spirit of this new localism is derived from self-identity, democratic participation, and cooperation.

While there is not, as yet, a cohesive formulation of concepts or rules of practice, we may be witnessing, through the writings in this text, the beginning of a new paradigm in the social sciences. This new field transcends traditional academic boundaries such as sociology, political science, and economics. It is hoped that these chapters will contribute to a coherent framework for guiding future research as well as strategy for attacking the problem of local industrial disintegration.

Traditional approaches to community economic development have emphasized securing help from outside the community—developers, new business, public relations specialists, and so forth. In too many instances, community development experts work behind the scenes, concocting corporate giveaway schemes such as tax credits or reduced charge public services, which serve only to cut local control over regional economic resources even more. Through such programs, outside interests are subsidized and unwittingly insulated further from local needs and regional realities. Thus a stable economic base becomes supplanted with a transient one in which outside groups come into the community, utilize what they need, and then move on, leaving shutdown plants and jobless residents.

In contrast, our emphasis has been on rebuilding the community from within, enriching local resources, and expanding the community economic base rather than playing into the national trend for low-paying high-tech and service-related industry. It is important to focus upon retaining existing jobs rather than simply seeking new work, to revitalize our older existing industry instead of passively assuming that the demise of smokestack America is inevitable. We need to move beyond an approach that operates by default to a strategy that fosters industrial entrepreneurship.

An underlying theme that runs throughout this book is the extraordinary resourcefulness of essentially ordinary people in combating the socioeconomic disintegration of their communities. Workers, managers, union leaders, local government officials, academics—all seem to be capable of bringing forth an impressive array of native skills for assessing problems and addressing the realities of absentee ownership, shifting capital strategies and industrial decline at the local level. We find it striking that the creation of area labor-management committees and conversion to employee ownership efforts have emerged in diverse areas and quite independently across the United States.

DECLINE TO INDUSTRIAL REVITALIZATION

Figure 16.1 illustrates the process whereby communities shift from local economic self-sufficiency to eventual industrial crisis. As local industry becomes increasingly linked to larger and more distant entities, the concerns of both local workers and the community become less and less a factor to be considered in corporate decision making. Under closely held or local family ownership it was once impossible to ignore the feelings and needs of the community and one's employees, but under absentee ownership it is perceived as more of an irritant and something to be avoided. As a firm becomes part of a broad and far-reaching conglomerate network of national or global investments, employees at individual plants and the communities in which those plants are situated are merely represented by financial ratios. With an ever-increasing emphasis on short-term, high-yield returns on investment, enterprises

Local Industrial
Ownership

↓

Absentee Ownership◄──────────────────┐
Control   │
                                     │
↓   │
                                     │
Conglomerate Ownership   │
Control   │
                                     │
↓   │
                                     │
Industrial   │
Decline   │
                                     │
↓   │
                                     │
Crisis   │
Community Choice   │
        ╱   ╲   │
      ╱   ╲   │
    ↓   ↘   │
Industrial   Industrial ─────┘
Revitalization   Promotion

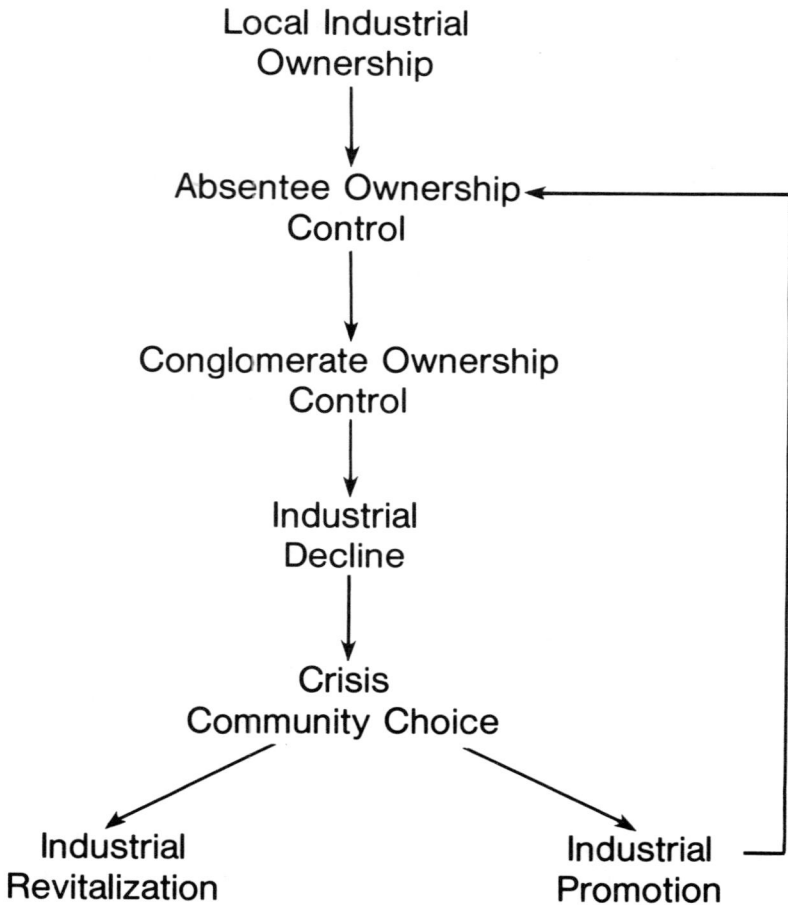

- Rebirth of Local Ownership
- Shared Influence Over Decisions

Figure 16.1

that once thrived and produced valuable and useful products fall into disarray and eventually are closed down when they lose financial favor. As this process transpires on an aggregate basis at the community level, gradual decline becomes a matter of economic crisis.

When faced with local economic strain, communities in the United States have a choice of either opting for the traditional "industrial promotion" strategy or, as some of the communities cited in this volume have, focusing on rebuilding their existing industrial bases. The problem with the first strategy is that it serves only to deepen local dependence upon outside corporate interests, and often results in the squandering of precious resources on what are essentially wild-goose chases. Tremendous tax and service giveaways frequently are required in order to "beat out" competitor communities, and in the end the bill must be paid by existing businesses and local citizens. Eventually, when industrial promotion is the sole or central strategy, the old vicious cycle of decline and crisis can and does ultimately return as other investments and new horizons offer better short-term yield.

In contrast, when a revitalization strategy is employed as the central thrust, the proportion of locally based ownership and leadership begins to increase, and again it becomes necessary for these groups, whether employees or local entrepreneurs, to include the concerns of the community and their employees in company decision making. And when a community-based labor-management committee serves as the central vehicle for change, absentee interests even find themselves having to share power with employees and local communities as they work together to solve problems and improve industrial performance.

NEW COMMUNITY INVENTIONS

The industrial revitalization efforts that have emerged in the country through both area labor-management committees and employee-owned firms all possess a uniquely American quality and, when successful, have relied heavily upon grass-roots volunteer action. In spite of the clear link with the distinctive U.S. preference for local autonomy and self-determination, we believe that these ventures represent a new adaptation of old values to a new and significant economic and social problem. Indeed, area labor-management committees and employee ownership can rightly be viewed as *new community or social inventions.* A social invention is defined here as any of the following:

- a new element in organizational structure or interorganizational relations
- new sets of procedures for shaping human interactions and activities and the relations of people to the natural and social environment
- a new policy in action
- a new role or set of roles

It is not important whether or not the social invention is "new" in some historical, factual sense. The important point is that the ideas that form the basis for the invention are new to the people involved in developing and applying them. The application also suggests newness in that the ideas are combined in the social, economic, and cultural dimensions of the particular situation.

The cases and experiences cited throughout this book amply illustrate the capacity of indigenous groups to invent new community approaches for coping with regional problems. Whether it be the Buffalo-Erie Labor-Management Council or the Hyatt Clark Industries buyout, local residents essentially banded together, more or less autonomously creating new structures of economic stability and industrial democracy. This was accomplished largely without any direct outside influence.

In an era of budget cuts coming from Washington, D.C., and a tightening of federal purse strings, we may see a growing number of grass-roots tactics being invented by groups and communities who have faced the fact that help will not come from outside. Instead of solutions from above, the shift is toward a bottom-up change strategy. This new self-help approach to economics may become a dominant wave of the future in community economic development. Clearly, labor-management committees and worker buyouts will remain important catalysts in the search for an effective community approach to revitalization.

A DIFFERENT MODEL
OF COMMUNITY INTERVENTION

In contrast to social inventions that emerge from within a community, *interventions* occur from outside. Here we are talking about the contributions and roles played by external resources to the community. In this volume we have attempted to illustrate a new model for government and expert intervention.

The traditional approach to community intervention has usually been led by the social scientist or consultant who comes into the situation with a packaged program to be installed in the community. The difficulty with such a method is that there is frequently a lack of fit between actual local needs and the outsider's solution. Often, the community rejects the external resource because of rigidity, the selling of "canned" solutions, and an emphasis on implementation rather than thoughtful, collective design.

What we see in the major cases described earlier in the book, such as the Jamestown Area Labor-Management Committee or worker-owned Rath Packing Company, is a different kind of outside intervention. In these instances, rather than outside experts entering the organization with a tight, preconceived answer to local economic strain and stress, the behavioral scientists used a different methodology.

This approach might best be described as a form of participative action research. Through observation, interviewing, and in-depth analysis of local culture and history, a rough map is created of the community and firm's social and economic territory. These patterns are tested out with residents for clarity and validity, reshaped as necessary, and further refined.

Instead of a packaged solution, a climate of collaboration is created between researchers and community leaders. Potential solutions are debated, tested, and more fully developed. The outsiders' role is to work in conjunction with, not *on,* local residents and workers. Local groups are not passive receptacles for receiving outside "interventionist" expertise, but co-creators of concepts, methods, and new economic structures.

For instance, in the case of the Muskegon Area Labor-Management Committee, consultants and leaders of local firms and unions jointly formulated a definition of problems and potential strategies. In the Hyatt case, expertise grew out of contacts with the UAW union local, not management. Many traditional development processes to improve business only focus on improving profits. The cases described in this book point to the need to create a more just society as well. ALMCs and worker ownership both seem to be strategies for accomplishing critical collective research that can lead to new institutions and/or the structural transformation of older, existing enterprises.

CONCLUSION

Perhaps as fitting a way as any to end this volume is to suggest where this newly emerging field of grass-roots reindustrialization should go from here. Previous chapters by Cutcher-Gershenfeld and Blasi articulate some new policy issues that need to be addressed. To these we would add the importance of research into the impact of mergers and acquisitions on employment. What kinds of industry tend to have a good track record when it comes to jobs before and after acquisitions? New research needs to occur on the effects of absentee ownership on local communities.

On the practical side, we strongly feel the need for funding that is oriented toward the development of new forms of industrial democracy. Fledgling efforts, such as those described in this book, need to be systematically investigated over the long-term as labor and management groups attempt to adjust to new roles and institutional relationships.

Finally, we would urge that various support mechanisms be created—policies, finances, support services—that are consistent with this new model of self-help economics and participatory democracy. The future, it appears, is going to require further refinement of efforts to forge new relationships between small companies and their employees to help each other. Indeed, helping people to help themselves may be a new theme in America over the coming decade.

INDEX

Absentee owners, 78-84, 184-186, 299-300
Action research, 107, 125-130, 136, 209-210, 303
Area labor-management committees, 88-90, 100-117, 245-257; outcomes, 113-116, 122, 131-133, 147-159, 173-177, 246; structure, 88, 89, 102-106, 154, 165; *Cases:* Buffalo, 161-178; Jamestown, 98, 141-159; Muskegon, 98, 121-138; Toledo, 88-89, 97

Bankruptcy, 146, 188, 201, 218
Behavioral science, 137-138, 298
Boards of directors, 94, 206-208, 232, 281
Bureaucracy, 228

Capital, 17, 31, 39, 53, 64-73, 185
Capital shift theory, 68-73
Codetermination, 229-233
Collective bargaining, 108, 224, 226-228
Communication, 109, 133, 150, 235
Community development, 23, 131-138, 298
Community development corporations, 18-20, 24-25
Conflicts and disputes, 134, 143-144
Conglomerates, 17, 31, 33, 37, 45, 182-185
Control, 37-39
Cooperatives, worker, 264-267
Corporate mergers, 46-47, 182, 304

Democracy, workplace, 208-213, 229-234

Economic Development Administration, 17, 41, 131, 147, 201, 208, 224, 270
Economic efficiency, 191-194
Education. See Training
Entrepreneurship, 27, 292-295

ERISA, 285-292
ESOPs (Employee Stock Ownership Plans), 186, 195-200, 207, 218, 223, 230, 264-267, 279-292

Federal government policy, 18, 254-258, 285-293
Federal Mediation and Conciliation Service, 19, 112-113, 165, 249
Financing: Labor-management committees, 111-113, 177-178, 247-251; worker buyouts, 223-224, 261-274

Grievances, 233

Industrial decline, 78-94, 121-122, 184, 201-202
Industrial democracy, 11-12, 159, 231, 304
Industrial development, 134-143
Industrial Development Agency, 19, 23-24
Industrial relations, 107-111, 209, 226-227, 252-254
Innovation, 294
Investment, 16, 31, 52, 84

Job creation, 16, 22, 158, 182-183, 208
Job preservation, 21, 50, 147, 157-158
Job Training Partnership Act, 247, 249

Labor-management cooperation, 21, 141-159, 168-177, 207-212, 287
Labor policy, 245-258, 285-293
Leadership, 105-107, 154
Legislation, 248, 254, 283
Lenders: banks, 115, 267; government, 17, 201-203, 271
Long, Russell, 195, 291
Lundine, Stanley, 23, 26, 145, 151, 248, 254

ABOUT THE CONTRIBUTORS

ROBERT W. AHERN is Executive Director of the Buffalo-Erie County Labor-Management Council, where for many years he has facilitated the implementation and expansion of in-plant and areawide programs to encourage union-management cooperation in western New York.

JOSEPH BLASI is Director of Kibbutz Studies, Harvard University, and a Lecturer in social studies. Previously on the staff of Congressman Peter Kostmayer, he has been a social policy adviser to several legislative committees in Washington, D.C., including that of the Small Business Administration. He is the author of *The Communal Experience of the Kibbutz*.

BARRY BLUESTONE is Professor of Economics, Boston College, where he teaches labor economics. He has written extensively on the problems of plant closings and is coauthor of *Capital and Communities* and *The Deindustrialization of America*.

JOEL CUTCHER-GERSHENFELD is currently a Research Associate at the Sloan School of Management, Massachusetts Institute of Technology. For several years he was Director of Communication and Research at the Michigan Quality of Work Life Council. He was the founding editor of *The Work Life Review* and has consulted with a number of ALMCs.

BENNETT HARRISON is coauthor of *Capital and Communities* and *The Deindustrialization of America*, two volumes on plant shutdowns and unemployment. He is a Professor of Political Economy and Planning, Massachusetts Institute of Technology.

STAUGHTON LYND is a former Professor of History at Yale University who became an attorney and joined the struggle of steelworkers in the Youngstown, Ohio, area. He has served as general counsel for various unions attempting to block plant closings, an experience documented in *The Fight Against Shutdowns*. Presently he is an attorney with Legal Services.

CHRISTOPHER MEEK directed the research program of the Jamestown Area Labor-Management Committee for several years. He taught at Boston College before going to Brigham Young University, where he is an Assistant Professor in the School of Management. A coauthor of *Worker Participation and Ownership*, he is working on a number of projects involving employee buyouts and labor-management cooperative problem solving.

DAVID MOBERG has a Ph.D. in anthropology from the University of Chicago, where he completed a dissertation on automobiles and working-class consciousness. He is currently a senior editor of *In These Times*, where he specializes in covering major issues facing the labor movement in the United States.

COREY ROSEN founded and serves as Executive Director of the National Center for Employee Ownership in Arlington, Virginia. A former staff member of the U.S. Senate Finance Committee, he directs a number of projects on the viability of employee ownership. His empirical analysis of the national movement toward employee ownership will soon be published.

WILLIAM FOOTE WHYTE is Professor Emeritus at the New York State School of Industrial and Labor Relations, Cornell University, where he now directs the school's Programs for Employment and Workplace Systems. Past President of the American Sociological Association, he is the author of many articles and books, including the early classic, *Street Corner Society*. His most recent volume is *Learning from the Field*.

WARNER WOODWORTH is Associate Professor of Organizational Behavior at Brigham Young University, where he teaches courses on industrial democracy and organizational change. A past visiting scholar at the International Labor Office in Geneva, Switzerland, and the University of Rio de Janeiro, he is completing a book on worker ownership and labor in the boardroom.

NOTES

NOTES

NOTES

NOTES